Lecture Notes in Logistics

More information about this series at http://www.springer.com/series/11220

Michael ten Hompel · Jakob Rehof
Oliver Wolf
Editors

Cloud Computing for Logistics

Springer

Editors
Michael ten Hompel
Fraunhofer-Institut für Materialfluss und Logistik IML
Dortmund
Germany

Oliver Wolf
Fraunhofer-Institut für Materialfluss und Logistik IML
Dortmund
Germany

Jakob Rehof
Fraunhofer-Institut für Software- und Systemtechnik ISST
Dortmund
Germany

ISSN 2194-8917 ISSN 2194-8925 (electronic)
Lecture Notes in Logistics
ISBN 978-3-319-13403-1 ISBN 978-3-319-13404-8 (eBook)
DOI 10.1007/978-3-319-13404-8

Library of Congress Control Number: 2014956490

Springer Cham Heidelberg New York Dordrecht London

Printed on acid-free paper

Springer International Publishing AG Switzerland is part of Springer Science+Business Media (www.springer.com)

Preface

Logistics is currently in a complexity trap: Globalization is constantly increasing, the degree of networking is growing exponentially, "batch size one" is already a reality. The volume of logistics data is increasing by a factor of 1,000 per decade. At the same time, the stability of logistics systems and processes is decreasing. Therefore, there is no doubt about the necessity of flexibilization of the classic Supply Chain Management. Cloud Computing is now helping to cut the Gordian knot: The Cloud as a system environment has the potential to design complexity to a certain extent controllably and processes more flexibly. It is about far more than new applications and markets. It is about the fundamental redesign of logistics systems. Cloud Computing has the ability to vitally change the kind of controlling and organizing of these systems.

The Fraunhofer innovation cluster "Cloud Computing for Logistics" is an extraordinary example to show the power that can be developed by logistics and informatics as drivers of innovations. The Logistics Mall with several awards in the meantime created a market place and an infrastructure that help providers and users to test the new possibilities of Cloud Computing in a practice-oriented way. For its users the Logistics Mall makes sure that Cloud Computing in the consciousness of logistics is more than a "cloud". Cloud Computing is understood as solution pushing forward logistics companies.

Logistics is surely one of the most important and most innovative application industries for information technologies (IT): it has internalized and pushed forward the topic of Cloud Computing because it realized very early the chances connected to it. However, the potential of Cloud Computing is by far not yet exploited as clearly pointed out in the position paper of Bundesvereinigung Logistik (BVL) e.V. "Logistics and IT as innovation drivers for the business location Germany—The new leading role of logistics in information technology". The Fraunhofer Institutes for Material Flow and Logistics (IML) and for Software and System Technology (ISST) meet these challenges and have bundled their competencies for this purpose in the Fraunhofer Innovation Centre for Logistics and IT (FILIT) since the middle of 2014.

Sponsored by:
Ministry of Innovation, Science and Research of the State of North Rhine-Westphalia
Fraunhofer Society (Fraunhofer-Gesellschaft zur Förderung der angewandten Forschung e.V.)

December 2014

Michael ten Hompel
Jakob Rehof
Oliver Wolf

Contents

Logistics Software Systems and Functions: An Overview of ERP, WMS, TMS and SCM Systems

Andreas Nettsträter, Tim Geißen, Markus Witthaut, Dietmar Ebel and Jens Schoneboom

Abstract Software systems are a crucial part in current logistics processes. The software market for Enterprise Resource Planning, Warehouse Management, Transport Planning and Supply Chain Management systems and the systems themselves are complex. The following article gives an overview about definitions and typical functions of these systems as well as a short introduction to current logistics and technological requirements.

Keywords Logistics software · WMS · ERP · TMS · SCM · Enterprise resource planning · Warehouse management · Transport planning · Supply chain management

1 Introduction

Logistics analyses and models the flow of goods, people and information through nearly all economic systems and provides recommendations for their design and implementation. The main focus in research is therefore on configuration,

---

A. Nettsträter (✉) · T. Geißen · M. Witthaut · D. Ebel · J. Schoneboom
Fraunhofer Institute for Material Flow and Logistics, Dortmund, Germany
e-mail: andreas.nettstraeter@iml.fraunhofer.de

T. Geißen
e-mail: tim.geissen@iml.fraunhofer.de

M. Witthaut
e-mail: markus.witthaut@iml.fraunhofer.de

D. Ebel
e-mail: dietmar.ebel@iml.fraunhofer.de

J. Schoneboom
e-mail: jens.schoneboom@iml.fraunhofer.de

M. ten Hompel et al. (eds.), *Cloud Computing for Logistics*,
Lecture Notes in Logistics, DOI 10.1007/978-3-319-13404-8_1

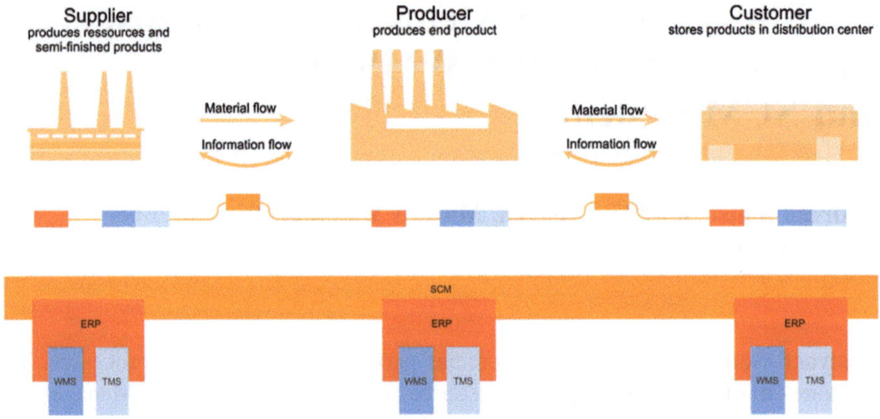

Fig. 1 Connection and information flow between ERP, WMS, TMS and SCM

organization, control and regulation of networks and flows with the objective to optimize these with economic, environmental and social objectives in mind.

Logistics software systems are defined within this paper as systems (meaning software and data) to support the above mentioned logistics tasks, including adequate planning and forecasting tools along the entire logistics supply chain. This includes all logistics areas from inhouse logistics over transport logistics to supply chain management. Typical software systems in these areas are Enterprise Resource Planning Systems (ERP), Warehouse Management Systems (WMS), Transportation Management Systems (TMS) and Supply Chain Management systems (SCM).

The general connection and information flow between ERP, WMS, TMS and SCM systems is shown in Fig. 1. SCM systems are not limited to a single company, but used within cross company supply networks. The ERP systems are usually company based just like WMS and TMS. The differences between SCM and ERP systems become more and more blurred. Many of today's ERP systems support multi-site capabilities and offer cross-site planning and processing functions and can therefore be used for the management of supply chains.

2 Definition of Systems

2.1 ERP

In the 1990s ERP systems with a large range of functionality for several company departments were mainly used in large companies. Small and midsize companies worked with expert systems for resource planning, production (PPS), Warehouse Management (WMS) and financial accounting. Fully integrated ERP systems became affordable for small and midsize businesses with decreasing cost for

Fig. 2 Key and extended functionality of typical ERP systems

hardware and software. In recent years ERP systems gained a more and more functional expansion and many former separated expert systems were integrated into the ERP systems. An ERP system today is a tool for comprehensive planning, coordination and management of companywide tasks. The advantage here lies in the efficient use of all existing resources of a company (e.g. capital, human resources and information). Apart from logistic applications (such as inventory management or disposition) ERP systems offer tools for almost all functions of a company, such as finance, accounting, controlling, manufacturing, research and development. Modern ERP systems must be able to adapt to dynamic business processes and they must change into flexible software solutions using open technical standards. Figure 2 shows the basic and extended functionality of standard ERP systems.

The market for ERP systems includes more than 200 vendors, but not everyone offers a self-developed ERP system. The market for ERP systems can be divided in the following categories:

- Software manufacturers develop and sell their own products and are responsible for all technological and functional properties.
- Resellers sell ERP systems as partners of software manufacturers to customers. Resellers are strong in market segments, where the software manufacturers do not distribute and sell their own ERP system.
- Suite vendors are also working with software manufacturers and enhance the functional scope of the ERP system by industry-specific modules or functions.

The advantage for the user is a consistent environment with all functions integrated.
- Implementation partners support the customer during the implementation of the ERP system with services.

There are also software manufacturers, resellers and suite vendors, who act as implementation partners at the same time.

2.2 WMS

Since the 1970s software systems for warehouses and inventory management are used to support logistic processes. Originally, these were systems to manage quantities and (storage or bin) locations as well as the relations in between them [1]. Additional functionalities could manage means of transport. Hence, warehouse management systems started as stock management systems.

After the major cost saving potentials of logistics in general and warehouses in particular were discovered, warehouse management systems no longer were systems to only manage the stock of a warehouse, but became integrated software systems with superior optimization and management functionalities.

Thus, warehouse management today describes the control, monitoring and optimization of complex warehousing and distribution systems. Next to the elementary functionalities of managing quantities and storage locations or controlling and scheduling the means of transport, comprehensive methods and instruments to supervise the system's status belong to the scope of a modern warehouse management system as well as a selection of operating and and optimization strategies. Consequently, the business of warehouse management systems is the operation and optimization of on-site warehousing systems.

Because customers constantly ask for further functionalities to realize the possibilities of cutting the costs for their warehouses, vendors of warehouse management systems constantly enlarge the scope of their software systems. As a consequence, warehouse management systems nowadays increasingly offer functionalities which have their origin in enterprise resource planning software, supply chain management software or transport management systems such as an entire order fulfilment, the support of all processes in between receiving and shipping and comprehensive information systems and control panels. Further functionalities which more and more often belong to the standard of a modern warehouse management system are the planning of tours and routes, the support of vendor managed inventories, or the support of billing and value added services in case of multi-client scenarios.

These core and additional functionalities (see also Fig. 3) are completed by optional subsystems (add-ons) for communication with the picker (e.g. Pick-by-Light, Put-to-Light, Pick-by-Voice controls) or systems to identify the products (e.g. RF readers, handhelds).

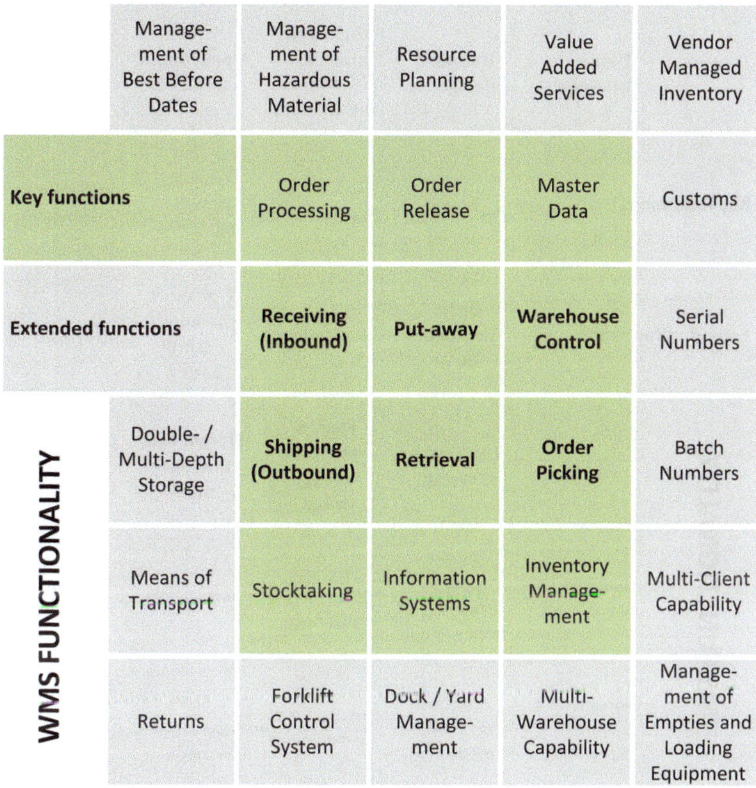

Fig. 3 Key and extended functionality of typical WMS

2.3 TMS

Before, IT-based transport management systems have been applied externally and transport related logistics was usually carried out with the aid of forwarder-customized software of the transport and logistics service providers. In case of more complex problems route-planning software was additionally applied. On the part of the shippers external logistics was covered by simple functional elements within ERP systems in the past.

Due to the increasing complexity of global and closely linked logistic chains at the end of the 1990s, the necessity of complex systems for the management of these complex transport chains with all their functional requirements began to develop. These systems are called transport management systems (TMS).

The resulting tasks cover the planning and optimization of procurement and distribution structures under the consideration of e.g. costs or time restrictions, the planning of multimodal transport chains or the optimization of the delivery transports as well as the control and monitoring of the realization of the resulting

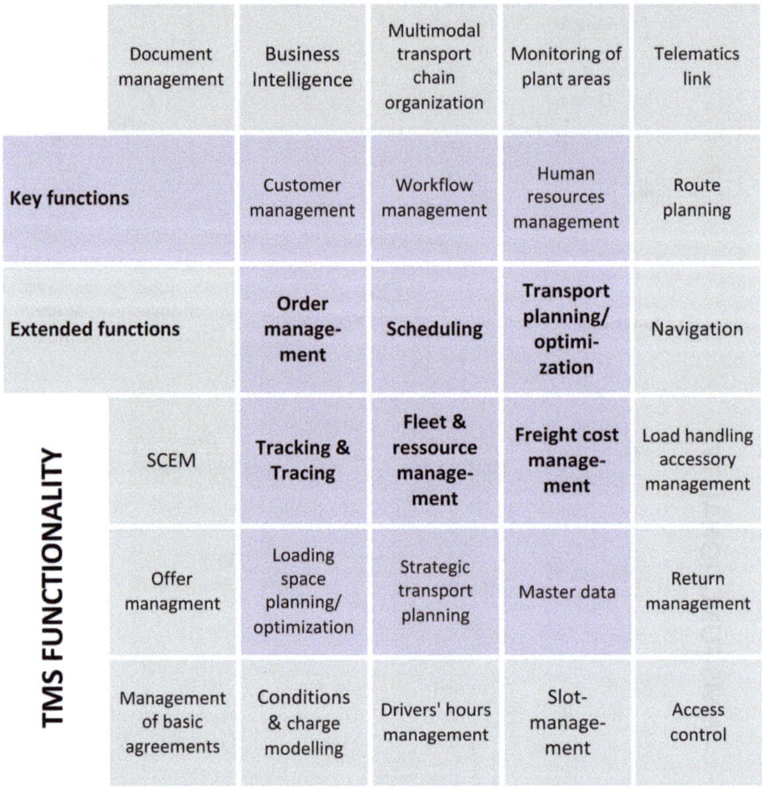

Fig. 4 Key and extended functionality of typical TMS

transport processes. The integration of the mobile units (transport means, load handling accessory equipment, etc.) by telematics is also an important element for control and monitoring.

TMS in general allow the planning, control and monitoring as well as the optimization of transport networks and logistic chains. Elementary functional fields of these systems are order management, scheduling, transport planning and optimization, tracking and tracing as well as fleet and resource management, see Fig. 4.

The market for transport management systems is already existing for several years. The enterprises providing transport management solutions can be divided in different groups:

Pure transport management software developers are providing their own products or as extensions in combination with other software elements. The reason could be found in a specialization of the software solution for a specific industry on the one hand. On the other hand it could also be found in the size of the developing enterprise. Therefore, the developing enterprises are cooperating with providers group of the sales partners selling the product to definite customers.

The sales partners are more service-orientated and solely focusing the sale of transport management solutions. Furthermore, there the so-called resellers who are also providing consultancy in addition to the sale of software. This allows the group to provide a wide range of different products—from the pure solution for road up to the combination of the transport means—respectively with a different range of functions—the reason for being very flexible and the ability to offer a wide range.

The providers can be described as a combination of the types mentioned above. Those are often medium-sized or big enterprises having an own development division as well as a well-organized sales division. For TMS this group is the focus, it is developing and implementing its software solutions directly and additionally making customer consulting.

Also the provided product range and the functional spectrum are often more versatile at big providers than at small enterprises. Furthermore, this group of providers is able to develop further software solutions beyond the system interface —e.g. ERP and WMS—thus also able to provide a complete package for bigger customers. Until now the number of enterprises also providing TMS in addition to ERP and WMS is limited. However, considering current extensions some of the ERP providers are increasingly moving towards TMS. In most cases an increasing range of TMS functions causes significantly increasing costs of the total system.

Due to the complexity of the products offered at the market both small and big enterprises will find a suitable offer of transport management systems.

In transport-intensive industries the application of software solutions is of significant importance. In the past years the demand for transport solutions and thus for transport management systems has been considerably increasing. This shows that the application of TMS is nearly unavoidable. In this respect the objectives of the users are scarcely differing. High saving potentials of the resources as well as a cost-optimal and efficient order management are more and more the goal. Differences in the requirements specification are mainly to be found in the size of the enterprise and the related horizon of the order spectrum.

The key competences of the TMS providers cover the planning and optimization of procurement and distribution structures with minimum costs which could be both, economic and ecologic ones. The systems, however, have also differences in view of the functional range and the technological base. Especially bigger providers show the trend of offering more strategic planning elements in their software. There are also transsectoral functions—as for example freight cost management or planning—provided with only few systems.

The complexity of transport management systems is often based in the integration of manifold frame conditions and restrictions—be that a legal regulation or the versatile and different demands of the customers for these systems. This and the worldwide opening of the transport networks are also considerably promoting the competitive pressure on part of the TMS -providers so that those are especially forced to continuously develop their systems.

The market development of the past years therefore shows the focus on constantly new requirements that the transport managements systems have to fulfill. The recent past indicated a trend in the modeling of multimodal transports. Even for complex European-wide or global transport chains with different transport means schedulers and planners are expecting a support for the planning and control, similarly to road-related transports being used to. In this context the ecologic assessment of especially multimodal transports and logistic chains also plays an important role.

2.4 SCM

Supply Chain Management (SCM) deals with inter-organisational production and logistics networks. The notion supply chain is somehow misleading because SCM addresses not only inbound logistics but the "business activities associated with all phases of satisfying a customer's demand" [2]. Furthermore, instead of single supply and/or distribution chains SCM covers arbitrary networks. Hence, the notion Value Net Management would be more suitable but is not used in the field. Similarly, the organisational coverage of SCM is differently defined in the field and the scientific literature. A point of view is that SCM integrates all logistical planning and execution tasks for sourcing, manufacturing, and delivery of a single plant's products. Another extreme opinion especially common in academic papers encompasses all sourcing, transport, manufacturing and distribution activities in network of separate enterprises spanning from raw material supplier to the final customer and including return and after-sales services as well.

The different SCM conceptions lead to a variety of definitions for SCM software. An extreme position is that SCM has to cover all logistical planning and execution task in cross-enterprise production and logistics networks. However, there is no such SCM software solution available on the SCM software market. Even though there are software vendors offering solutions for the entire planning and execution tasks, these systems cannot be applied across enterprise boundaries. For a pragmatic approach we recommend a conception of SCM as the design, planning and execution of both intra- and inter-organisational material and information flows. Consequently, we propose a task model encompassing all activities with IT support for the management of production and logistics networks (see Fig. 5). The software systems can be further structured into three main groups. The first group serves for the design of networks—Supply Chain Design (SCD). SCD-Systems cover issues like location planning including, transport system design and the dimensioning of warehouse in respect to customers' demand and production capacities. Supply Chain Planning (SCP) includes planning tasks for a given network. Based on customers' demand, overall sourcing, manufacturing and distribution plans are generated. Furthermore, some SCP-Systems offer features for the collaborative demand and capacity planning with customers and suppliers. Available-to-Promise functionality is used for due dating customer enquiries on a supply chain level. The last SCM-functionality group is called Supply Chain Execution. With the exception of supply chain event management, SCE-functions overlap with the functionality of ERP, TMS and WMS software (see Fig. 5). An initial version of this SCM task

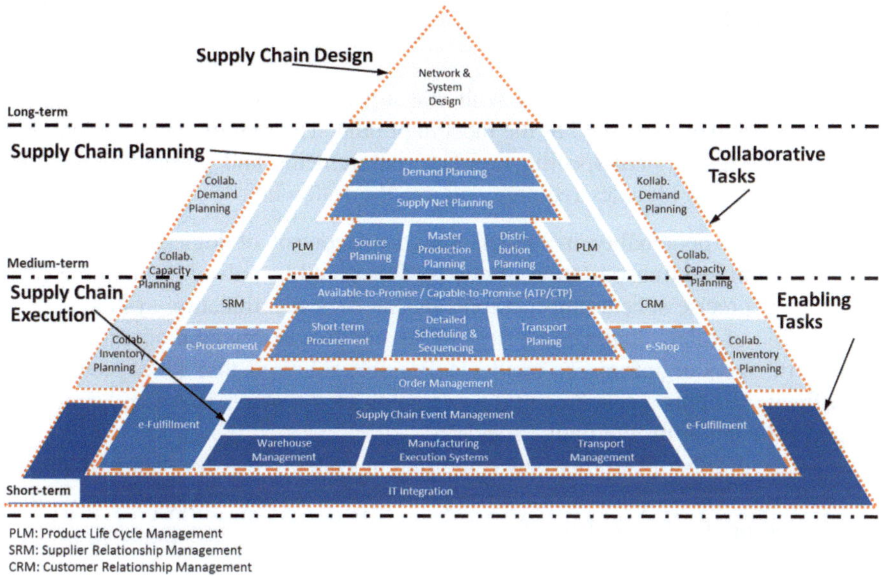

Fig. 5 Key and extended functionality of typical SCM systems

model was developed especially for a comparative market study on supply chain management software in 1998/1999. A major revision was made in 2003. The resulting model comprises the typical modules of SCM software and is further based on the internationally established Supply Chain Operations Reference-model (SCOR).

3 System Requirements

The current evolution of logistics system shows that modern information technology will be more and more coupled to all fields of logistics. Only through this close coupling the complexity of logistics can be handled and already today there are just a few fields of logistics left, which are able to work efficiently without the use of IT systems. The software solutions used are as numerous as are their heterogeneous areas of applications. However, there are still many barriers both, from the user's perspective as well as from the perspective of the software provider, contrary to the use of IT in logistics. The next section will give a brief overview about user requirements to logistics software systems. Additionally, many technological trends will find their way into logistics IT systems which are covered in Sect. 3.2.

3.1 Users Requirements

Based on a survey made last year with 208 logistics software users and providers [3] ten requirements emerged as the most important. The requirements were divided into two groups, system-specific logistics requirements and global technological requirements. The logistics requirements are the following:

- Increase of punctuality in the planning and control of logistic processes for production and transportation of goods. This could be solved through the usage of more precise scheduling methods and modern identification systems.
- Creation of robust logistics systems which are able to continue operation also in case of errors or unforeseen incidents. Possible solutions that are expected of the software systems are, sufficient dimensioning of resources, integrated simulation and calculation of alternative scenarios.
- Increase of interoperability between different systems, e.g. for data or document exchange. This requirement is closely coupled to the technological requirement of open or standardized interfaces.

3.2 Technological Requirements

Technological requirements often mentioned in the survey for future logistics IT systems are briefly described below:

- Flexible: The system must be expandable for future changes and new (currently unknown) requirements. The ability to respond to unforeseen events is known as the changeability of systems.
- Individual: Custom requirements and additions can easily be implemented.
- On-Demand: The software should be available directly, without a long-term implementation phase or the need for additional IT infrastructure.
- Open/standardized interfaces: The software should use public and open interfaces for information exchange so that other systems can be connected without major integration effort.
- Low cost: The capital and operating costs must be minimized.
- Intuitive HMI: The graphical human-machine-interface should be self-explanatory and understandable.
- Usable on mobile devices: The software should be accessible also from outside the company via mobile devices, like smartphones or tablets.

These requirements make it inevitable for the providers of logistics IT to keep an eye on current technological trends which can offer possible solutions to meet these requirements.

References

1. ten Hompel, M., Schmidt, T.: Warehouse Management. Organisation und Steuerung von Lager- und Kommissioniersystemen. Springer, Berlin (2008)
2. Supply Chain Council: SCOR 10.0 Model Reference. Supply Chain Council, Cypress (2010)
3. ten Hompel, M. (ed.): IT in der Logistik. DVV Media Group, Hamburg (2012)

Logistics Mall—A Cloud Platform for Logistics

Damian Daniluk and Bernhard Holtkamp

Abstract Common characteristics of all logistics processes are individuality and dynamically changing requirements of the customers' business. Cloud computing enables new business models to provide highly individual IT services that fit the needs of logistics customers. After outlining logistics specific cloud service requirements and the results of a study about the acceptance of cloud computing in logistics domain this paper presents the Logistics Mall, an approach for a domain specific cloud platform for the trading and usage of logistics IT services and logistics processes.

Keywords Logistics mall · Cloud computing · Cloud platform · Business process as a service

1 Introduction

In 2011, the overall global logistics market represented a total volume of approximately 981 Billion € [1]. According to BVL (recognized german logistics institution), in Germany logistics is the third largest economy branch with an market volume of approximately 235 Billion € in 2014 and a work force of 2.9 Million. The German logistics market, as typical for the logistics domain, is characterized through small and medium size enterprises with little or no IT capacities and competences besides operating their own IT resources.

D. Daniluk (✉)
Fraunhofer Institute for Material Flow and Logistics IML,
Joseph-Von-Fraunhofer-Str. 2-4, 44227 Dortmund, Germany
e-mail: Damian.Daniluk@iml.fraunhofer.de

B. Holtkamp
Fraunhofer Institute for Software and Systems Engineering ISST,
Emil-Figge-Str. 91, 44227 Dortmund, Germany
e-mail: Bernhard.Holtkamp@isst.fraunhofer.de

M. ten Hompel et al. (eds.), *Cloud Computing for Logistics*,
Lecture Notes in Logistics, DOI 10.1007/978-3-319-13404-8_2

Trade and industry consider logistics as a cost factor and as a factor of competitiveness at the same time. As a consequence there is a growing trend for outsourcing and contract logistics to benefit from scale effects and from synergies. This way the market for logistics services has developed from classical transport—transshipment—warehousing services towards a growing market of more individual services which are offered by specialized logistics service providers. There is also a trend this services become more complex increasingly. One of the reasons for this is the development in the e-commerce sector. In Germany, purchases over the internet result in more than 100 million additional packages per year. Each of these packages is ordered individually and on demand, picked, packed, transported, distributed and delivered. At the same time, the number of items increases exponentially. This means that customers of logistics service providers today require individualized logistics services with:

- a flexible and broad service spectrum,
- individual logistics processes and value-added services,
- transparency of costs and performance,
- short-term contracts.

These requirements increase the following problems, evident for many logistics service providers:

- It usually lacks sufficient investment funds (or the willingness to invest) for the expansion of IT.
- Often it lacks IT expertise and availability of human capacity to operate the necessary IT infrastructure adequately.
- The development of new IT components and their integration into the existing IT landscape is time-consuming.

Many logistics companies neither have adequate IT skills and capabilities, nor the capital needed to close the gap between requirements and the status quo in the logistics IT. In addition, there is growing pressure on the logistics sector through legislative changes. Moreover, street charges and the need for electronic customs clearance imply the use of new IT services, e. g. for control of driving and rest times. The same applies to the introduction of tracking and tracing for continuous condition monitoring of goods.

In this context cloud computing is a very interesting technology to face the outlined challenges. The rest of the paper is organized as follows. In this chapter the requirements of logistics on application domain specific cloud services are discussed. Section 2 summarizes the results of a study about the acceptance of cloud computing within the logistics domain in Germany carried out by the Fraunhofer Institute for Material Flow and Logistics IML. Section 3 provides an overview of the Logistics Mall development. The paper closes with a summary of results and an outlook to future work.

2 Cloud Computing in the Logistics Domain

Cloud computing is considered as a megatrend which will have an impact on the usage of information technology in all application domains. Forrester forecasts that the global market for cloud computing will grow from $40.7 billion in 2011 to more than $241 billion in 2020. The total size of the public cloud market will grow from $25.5 billion in 2011 to $159.3 billion in 2020 [2].

For a common understanding we adopt the working definition of NIST [3] that defines cloud computing as "a model for enabling convenient, on-demand network access to a shared pool of configurable computing resources (e.g., networks, servers, storage, applications, and services) that can be rapidly provisioned and released with minimal management effort or service provider interaction. This cloud model promotes availability and is composed of five essential characteristics, three service models, and four deployment models." The basic characteristics are on-demand self-service for consumers, broad network access, resource pooling to serve multiple consumers using a multi-tenant model, rapid elasticity of resources and metering capabilities for service provision. The identified service delivery models are software-as-a-service (SaaS) where consumers use a provider's application running on a cloud infrastructure, platform-as-a-service (PaaS) where a consumer can deploy an application on a cloud infrastructure using a provider's tools and platform, and infrastructure-as-a-service (IaaS) where a consumer can run arbitrary software on a provided cloud infrastructure.

Recently, big IT companies have invested many million dollars in the development of cloud technology and in the provisioning of commercial cloud offers. The span reaches from IaaS solutions like Amazon's Elastic Compute Cloud [4] to PaaS offers like Microsoft Windows Azure [5] and SaaS offers like Salesforce CRM [6]. A common characteristic of these offers is their generic nature. They all concentrate on broad usage scenarios but do not provide business relevant features. These cloud offers are called "horizontal clouds". In their report on the future of cloud computing [7] a European expert group identifies the provisioning of application area specific cloud services, also called "vertical clouds", as a significant business opportunity. The German excellence cluster EfficiencyCluster LogisticsRuhr (www.effizienzcluster.de) shares this view and has declared the development of a domain specific logistics cloud a strategic target.

2.1 *Requirements on Logistics Services Out of the Cloud*

Logistic applications often have specific characteristics that are implied by the requirements of the industry. By shifting logistics applications to the cloud, these characteristics have to be considered in order to provide a proper replacement for the existing customizable software solutions. The essential requirements that must be considered in the context of logistics specific cloud services include:

- Integration of customer-specific peripherals
- Realization of short response times of logistics software
- Ensuring data security and transparency
- Timely implementation of highly-customized business processes at low cost

Many of these requirements are not just technical in nature. However, they lead to technical requirements on cloud computing on closer inspection.

2.1.1 Integration of Customer-Specific Peripherals

Typical software systems used in intralogistics include warehouse management systems (WMS). In addition to the basic functions including the management of quantities, bin locations and transport orders a warehouse management systems also includes extensive means of controlling the system states and implements a variety of operating and optimization strategies. Typical peripherals that have to be interfaced with a WMS include [8]:

- Forklift and handheld terminals: These components are wireless computers for mobile applications such as for forklifts or for manual operation. They are used for timely and easy collection of data, for example as part of the picking of products.
- Barcode reader (infrared scanners) and transponder reader (RFID scanner): Scanners are wired (e.g. USB) or connected wirelessly to a PC. The captured data is processed by the WMS.
- (Label) printers: Printers serve in the context of WMS as a medium for the creation of goods receipt documents, picking lists or goods issue papers. Special label printers can write adhesive labels which are attached as forwarding labels on pallets.

2.2 Realization of Short Response Times

In logistic applications, compliance with upper limits of the response time of the software is very important. For example, if it takes too much time to update a user dialogue in applications like high performance picking, the employee cannot continue with his work (e.g. scanning of products) and is hindered in his work routine. This results in a reduction of productivity. To prevent such occurrences, cloud applications should be high-performant, i.e., the response time of the application for user input must be as low as possible so that user productivity is not affected. Particularly, the use of cloud computing may not lead to an obvious reduction in the performance of the user-frontend.

A particular challenge is the control of real time material flow systems. In the field of continuous conveyor systems guaranteed response times of less than 10 ms are standard. This can only be realized through the use of a material flow control systems on site, which may be coupled with applications in the cloud that are less time-critical. On the other hand autonomous designed conveyor modules increasingly allow a decoupling of real-time operations and thus the direct combination of cloud and material flow control.

2.2.1 Information and Data Security

Information and material flow in the logistics domain have reached a very high level of complexity. On the one hand this results from the variety of involved entities (suppliers, customers, means of transport, machinery, goods, IT systems, etc.). Another reason for this are multiple dependencies between the involved entities. Logistics is responsible for the planning, control and optimization of material and information flow that goes beyond company boundaries. As part of the cooperation between companies more and more short-term co-operations are formed, which depend on the core competencies of the companies. The tendency of this development is increasing.

The logistics of today is not imaginable without modern IT, wireless data transmission and mobile devices. Decentralization and particularly networking play an increasingly important role in logistics. This development leads to new requirements for existing security concepts. Especially in the logistics domain, the customers of logistics service providers have a legitimate, strong interest to protect their data (e.g. throughput data, sales data, and master data). Also, the use of mobile devices (GPS, telematics, RFID readers, PDAs, smartphones and laptops) is ubiquitous in logistics. Especially here it is important to develop concepts that enable the safe use of such mobile hardware in the cloud.

2.2.2 Individual Business Processes, Implementation Timeframes and Costs

In particular, the time factor blocks the development of flexible and innovative IT-based logistics solutions in several ways. The implementation time of today logistics systems is very long in proportion to the period of their utilization. Thus, business models and strategies in industry and commerce currently have a typical life cycle time of 2–3 years. The adequate modeling of (logistics) business processes often claims a realization time of 6−10 months. If additionally there has to be implemented a new, efficient IT solution for providing the required logistics services, then the implementation period extends to an average of 12 up to 24 months. The operating time of such strategic investments is an average of 3−5 years. After this time period normally the IT solution has to be adjusted because of new or changed requirements. This leads to an economically problematic relationship between planning and realization expenses and usage times [9].

This problem can be mitigated by IT components that can adapt to new requirements simply and quickly by dynamic combination to support more complex processes. For a cloud computing infrastructure this means that it must support the loose coupling of functional services in the form of autonomous IT services. A complete business process can be built up on this basis by choreography or orchestration of IT services [9].

2.3 Acceptance of Cloud Computing in Logistics

The acceptance of cloud computing in logistics was analyzed in the current market study "Cloud Computing for Logistics" of the Fraunhofer Institute for Material Flow and Logistics IML in Dortmund [9]. In this analysis providers of logistics IT services and potential users from the fields of logistics service providing, commerce and industry were interviewed. The goal was to find out the conditions under which the logistics and IT managers are willing to consider and use cloud computing approaches for critical logistics applications such as warehouse management systems (WMS).

The results of the study speak for themselves: the degree of acceptance of logistics solutions in the cloud is very high. Already today the majority (56 %) of corporate leaders can imagine to rent and run logistics software an external servers. This shows that the market is open-minded for the usage of logistics software from the internet and there exists a great potential for this form of software distribution.

The following section describes the "Logistics Mall", a platform that allows the purchase and usage of logistics software from the internet based on cloud computing.

3 Logistics Mall

3.1 Vision

Common characteristics of all logistics business processes are individuality and dynamically changing requirements of the customers' business. A process has to be adapted to every new or changed requirement. Consequently, the underlying IT infrastructure is also subject to these frequently changing requirements.

Today a typical logistics process runs in a heterogeneous IT environment. Such systems are mostly standard software like SAP, Oracle, or more specialized logistics software systems like warehouse management systems (WMS) or production planning systems (PPS). All these systems do not entirely meet the requirements of logistics customers regarding short contract duration, pay-per-use accounting and the provisioning of individual IT services. The majority of the features of standard software is rarely used or not used at all, but has to be paid for, due to a monolithic architecture. Customizing software to adapt to new or changed requirements involves high efforts, costs and risks. Nevertheless the solution's flexibility for upcoming adaptations is not increased.

The introduction of cloud computing is a new opportunity to deliver different IT services over the internet. The idea of this IT paradigm is the abstraction of the underlying software and hardware. Based on this, users do not have to manage and administrate the physical hardware or software they are using. An additional advantage is that software can be acquired on-demand and paid per use [10].

The concept of providing services over a cloud encompasses three different stakeholders. Basically these are the operator of the cloud computing environment (CCE) being responsible for its administration, providers offering their applications or IT services and customers purchasing and using the services [11].

Regarding the dynamically changing requirements cloud computing provides the opportunity for a logistics customer to rent IT services only as long as they are needed. Furthermore, using a CCE to provide and use services enables customers and logistic service providers to focus on their key business by outsourcing the IT infrastructure to the cloud. Both customers and logistic service providers do not have to establish and administrate an internal IT infrastructure and only require a connection to the internet for interacting with the cloud. Additionally, pay-per-use accounting is another advantage for the customers. For providers cloud computing is a new opportunity to gain greater market relevance and design new offers by connecting their products with services provided by other independent providers (according to the slogan "The whole thing is more than the sum of its components") [12, 13].

3.2 Concept

The main idea of cloud computing—"Anything as a Service"—can be adapted to source complete logistics business processes, designed by connecting single IT services using an appropriate tool that is executed in the browser. Therefore, an adequate process modeling methodology is necessary to offer logistics customers an opportunity to model individual processes themselves. This service delivery model is defined as Business Process as a Services (BPaaS). There are contributions that use BPaaS as abbreviation to name models that have the general idea of outsourcing of existing processes into the cloud [14]. Here, more specifically, the unique feature of this model is that both modeling and process execution are part of the cloud platform. This idea is the basis of the Logistics Mall, a development of the Fraunhofer Innovation Cluster "Cloud Computing for Logistics" [15], that focusses on the modeling and execution of processes that are built out of several IT services and that can be offered using a cloud platform. Most of the today available approaches are too complex to be used by logistics customers that have minor software engineering skills. On the one hand the vision of the Logistics Mall is to create a methodology that is capable to deliver business processes that can be deployed automatically for execution. On the other hand the logistician may not be overburdened by technical details.

The main idea of the Logistics Mall and the offering of BPaaS is visualized in Figs. 1 and 2. Today's monolithic software solutions are replaced by small, dedicated IT services of different service providers which can be combined to superior services that support the individual business processes of the logistician.

In addition to the offering of BPaaS solutions the Logistics Mall is also able to provide classic IT services corresponding to the Software as a Service (SaaS) cloud model. The main reason for this is to address as many software providers as possible to be able to offer IT systems that support a broad range of logistics processes. For an IT service that is offered in conjunction with BPaaS the technical

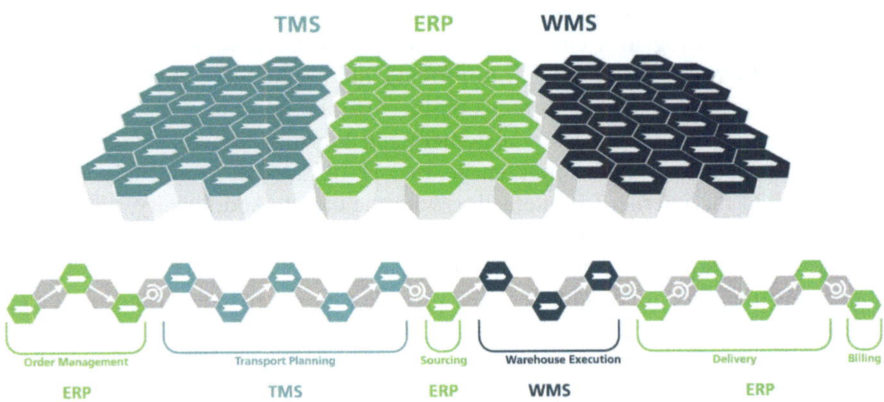

Fig. 1 Today's monolithic software solutions and their interaction within a business process

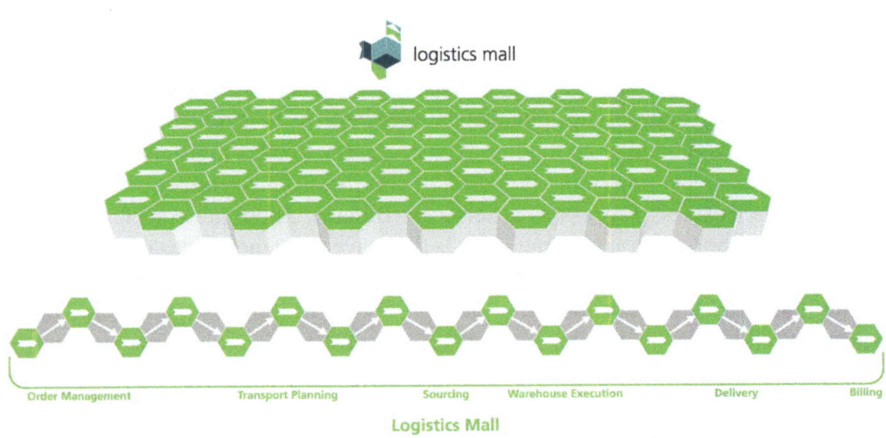

Fig. 2 The Logistics Mall as a cloud platform offers IT-services that can be combined to support individual logistics business processes

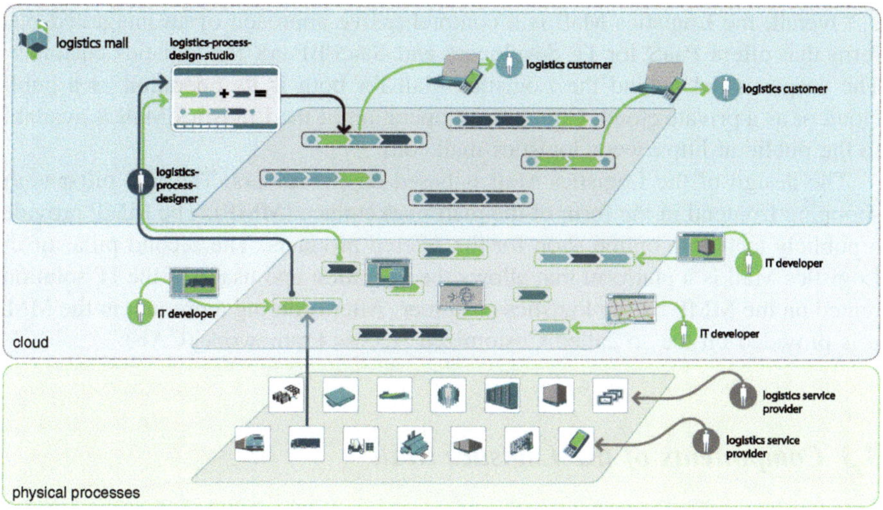

Fig. 3 Concept and stakeholders of the Logistics Mall

requirements are much higher than for an IT service that is offered only in as SaaS in the manner of ASP (application service providing). The reason for this is that BPaaS requires the IT service to use the same business object data model so that in the context of a modeled process IT services of different service providers have a common understanding of the data that is communicated between them. A detailed description of this data model and how it supports logistics business processes and reduces the business-IT gap is provided by [16]. Moreover the IT services in context of BPaaS should have a consistent look and feel of the GUI when they are used to support one business process. This requirements force service providers to undertake broad modifications on their existing IT services to be offered on the Logistics Mall. In contrast, providers have to do only slight modifications to their web-based IT services to offer them as SaaS products on the Logistics Mall platform (cf. Sect. 3.4).

Part of the concept of the Logistics Mall (Fig. 3) is that *IT developers* can offer and operate IT services in the cloud. Another target group is companies that have the function of integrators. As *logistics process designers* they combine IT services, which are compatible to be used in context of BPaaS, to more widespread IT solutions that support complete logistics business processes. The third group is the *logistics customers*, who order and consume the IT solutions offered by the Logistics Mall.

Conceptually the Logistics Mall also supports the integration of physical services. Such serviced could be offered on the Logistics Mall platform by *logistics service providers*. Examples of these include transport services, which encompass a transport of goods from a source to a destination.

Overall, the Logistics Mall is a comprehensive approach of an integrated platform that offers PaaS for IT developers and SaaS/BPaaS for logistics customers. The usage model behind the Logistics Mall for both is its operation as a public cloud or as a private cloud. Commercial operation of the Logistics Mall is available to the public at http://www.logistics-mall.com.

The design of the Logistics Mall is based on two pillars. The first pillar is the shopping frontend in the form of the mall marketplace (MMP). The MMP provides a publicly available online shop for the offered products. The second pillar of the Logistics Mall is a platform that allows the operation and usage of the IT solutions rented on the MMP by the logistics customer. After ordering a product in the MMP it is provided on the so called Customized Access Framework (CAF).

3.3 Components of the Logistics Mall

The following section describes some essential elements of the Logistics Mall from a functional viewpoint. An architecture overview about the Logistics Mall is presented in [17].

3.3.1 User Management

User management in the Logistics Mall is realized as a cross-component that is used at both the MMP and the CAF. Every user account of the Logistics Mall is assigned to an explicit company, in the wording of the Logistics Mall called client. A distinction is made between three types of clients, namely the operator, the providers and the customers. The operator has an integral access to the Logistics Mall through which he can access all the functions necessary for administration. Providers and customers get access to the Logistics Mall as part of a registration process. During this process, every company gets assigned a unique client number. Besides user name and password as user-specific identity information the unique client number is required for the MMP login. For the CAF login the client number is not necessary because the CAF is provided client-specific and is accessed via a unique ULR assigned to the client.

The user management of the Logistics Mall allows a client to create arbitrary other user accounts for users of his company and to assign permissions to each of these accounts. The permissions relate to functions of both MMP and CAF. For example, a user of type customer that has all possible permissions can access both MMP and CAF and use all available functions and rented IT services with their full functional range.

The assignment of permissions to users is done via roles. Each role aggregates a set of permissions. On the one hand roles may have been created in the user management of the Logistics Mall and then refer to functions that are provided in MMP and CAF. On the other hand roles may have been exported by the IT services the customer has rented and refer to permissions within the specific IT service. The user information and the roles of a logged in client can be requested by the IT services using a specific Logistics Mall interface. This concept called single sign on is realized across the entire Logistics Mall. A user only needs to login once and then can use all the features and IT services that he is allowed for by the permissions of his user account.

Thus, the user management built into the basic infrastructure of the Logistics Mall is a central element, which simplifies the usability of the Logistics Mall and reduces the organizational overhead.

3.3.2 Product Management

To capture new IT services in the Logistics Mall, there is a dialogue-based wizard, which queries all necessary product information from the IT service provider in a structured form. The registered provider can access the wizard on the MMP.

In the first step of the wizard, the general product information such as product name, version, classification and text, logos and flyers for product description are queried. In the second step the specification of the accounting model is done for the product. A distinction is made here between three accounting types: one-off costs, flat-rate costs (in regular intervals recurring costs) and transaction-based costs which support pay per use accounting. Transactions can be seen as clearly identifiable events that written by an IT service using a given logging interface of the Logistics Mall. Through the analysis of these logs a transaction-based billing is possible. A combination of these three accounting types is called pricing model. A pricing model may be assigned to the product itself or to any associated product option separately.

The specification of product options and associated pricing models is the third step of the dialog based wizard. Product options describe additional services that can be ordered by the customer together with the product. An example for such additional services is functional add-ons for the product. The product options can be aggregated into so called option groups. In addition, it can be specified whether a customer may choose only one option or several options from an option group as part of the ordering process.

In the fourth step of the wizard information about contractual design and accounting of the product is queried. These details include the minimum contract period and cancellation period. In the final fifth step, a preview of the product is presented. The provider can see how its product will be presented in the public MMP product catalog. The Logistics Mall workflow requires a product to be certified and released by the operator before it is presented in the MMP product catalog and can be rented by customers.

Generally, the product management of the Logistics Mall provides a comprehensive instrument with which the product and its associated accounting model can be described individually by the provider of an IT service.

3.3.3 Logistics Process Designer

The logistics process designer (LPD) is the central tool of the Logistics Mall that allows the modeling of IT solutions that support the individual business processes of a logistician. To be ready for business, the LPD supports a modeling methodology with the following characteristics:

- The LPD can be used by a logistics expert without detailed IT knowledge.
- A process consists of available IT services only. Modeling an arbitrary process just to notice it cannot be run by the IT services available is not practical.
- The LPD supports in its modeling process different functional granularity of IT services and the CCE. Forcing IT service providers to only implement services with specified, fixed granularity limits the distinction of IT services and hampers the growth of a broadly diversified range of IT services covering all ranges of functionality demanded by customers.
- The IT services within the modeled logistics process are exchangeable.
- On premise IT systems that are not operated in the cloud can be integrated into modeling by specific interfaces. By this, customers have the opportunity to migrate to the cloud step by step, still using existing IT systems and connecting them with IT services provided by the cloud.

A detailed description of the approach of the LPD is discussed in [18].

3.3.4 Reporting

Each client of the Logistics Mall gets access to a client-specific documents area. At adjustable, regular intervals, reports are stored that contain diverse dynamic data:

- Tickets: statistics about tickets, that have been sent using the helpdesk of the Logistics Mall
- Hardware utilization: details about utilized CPU time and memory utilization originated from IT service operation in the CAF
- Transactions: Data of the transactions the IT services in the CAF have written for pay per use accounting

Reports are customized specifically for the client types operator, provider and customer. For example, the transactions report for end customers contains the number of transactions logged for each rented product. For the provider the same report contains transaction numbers for each of his products and for each of his customers.

The reporting component of the Logistics Mall allows clients to trace data that occurs in operation and is relevant for business and accounting.

3.3.5 Helpdesk

The Logistics Mall incorporates an integrated ticketing system, which enables the processing of support requests. The 1st level support for end users is carried out by the operator of the Logistics Mall. With this the operator represents the first instance to which customers direct their support requests. If the processing of a request is not possible by the operator because the problem-solving requires a deeper understanding of an application, the operator forwards the request to the provider of the application. The provider thus takes the 2nd level support and can reply to requests of customers using the integrated ticketing system of the Logistics Mall.

Since the ticketing system belongs to the basic infrastructure of the Logistics Mall, users can access it from both MMP and CAF. This provides a platform that can be used to discuss and solve customer's problems.

3.4 Interfaces for Application Integration

To make applications available in the Logistics Mall, interfaces are provided for the application integration that must be implemented by the provider. The application integration is largely based on Web service interfaces, as this will ensure that the interfaces can be addressed through almost all modern programming languages. Subsequently the main interfaces of the Logistics Mall are shortly described. These are related to classic IT services corresponding to the Software as a Service (SaaS) cloud model. Interfaces of IT services, which are used in the context of BPaaS in the Logistics Mall, are described in [19].

- IT service GUI integration: two different interfaces are provided for the integration of IT service GUIs. Supported are already web-enabled applications. A simple integration is provided with the help of so-called inline frames. They allow that a browser window containing the IT service GUI can be shown as component of another browser window. In addition, application GUI can be integrated via portlet technology.
- Transaction-based accounting: to implement the transaction-based accounting for each application the relevant business transactions are queried within the product management wizard (cf. Sect. 3.3). During operation the application then reports its transactions through an interface to the Logistics Mall. The transactions are logged for accounting purposes.
- Connection of external peripherals: for reliable connection of the typical peripherals in logistics environment a special interface is provided by the Logistics Mall. This interface allows an application to establish a secure connection to the periphery.
- File storage: applications often write log files, reports or exchange data using a shared file. These files also must be stored in a cloud environment like the Logistics Mall. For this reason, Logistics Mall provides an interface through

which IT services can read, write and delete files and folders that are stored in the cloud.

- Error Handling: because applications and services are not operated locally or in the company-internal infrastructure, it is important that they can report errors. To integrate the error handling with support the Logistics Mall provides an interface to the ticketing system. Using this interface an IT service can report errors using a Web service or sending a simple e-mail to the ticketing system. In both cases error handling is integrated with the ticketing workflow.
- User and role management: a key requirement for the provisioning of services and applications from the Logistics Mall is that after a one-time authentication a user can call all its associated services and applications without repeated login procedure (Single Sign On). The Logistics Mall therefore provides an interface that can be used by IT services to be query for user and role information.

4 Conclusion

The Logistics Mall platform aims at providing a market place for logistics services and business processes together with a cloud-based access and execution environment. The currently Logistics Mall platform that is accessible under www.logistics-mall.com for the public supports the offering of IT services using the SaaS delivery model.

In the next step this platform will be extended towards support for the design and execution of workflows that combine logistics IT services and physical logistics activities using the BPaaS delivery model. With regard to the process execution a proof of concept implementation is being realized in form of a reference scenario. This reference scenario includes the execution of different types of business processes for warehousing within the intralogistics domain. The activities range from support for the incoming goods, the storage and picking to the shipping process and the outgoing goods. Part of the implementation is the integration of peripherals typically used in logistics, such as label printers and handheld scanners. The modeling and design methodology using the LPD is also demonstrated and validated by the reference scenario. The results obtained so far suggest that the proposed methodologies a viable, meaning that the proposed concept is suitable to be used in industrial applications.

References

1. Doll, A., Friebel, D., Rückriegel, M., Schwarzmüller, C.: Global logistics markets. http://www.rolandberger.at/media/pdf/Roland_Berger_Studie_Global_Logistics_Markets_fin_20140820.pdf (2014). Accessed 24 Nov 2014
2. Forrester Research: Sizing the cloud—a BT futures report. Understanding and quantifying the future of cloud computing. http://www.forrester.com/Sizing+The+Cloud/fulltext/-/E-RES58161 (2011). Accessed 08 April 2013

3. Mell, P., Grance, T.: The NIST definition of cloud computing. Working paper National Institute of Standards and Technology. http://www.nist.gov/itl/cloud/upload/cloud-def-v15.pdf (2009). Accessed 08 April 2013
4. Amazon Elastic Compute Cloud (Amazon EC2). http://aws.amazon.com/ec2/ (2013). Accessed 08 April 2013
5. Microsoft Windows Azure. http://www.microsoft.com/windowsazure (2013). Accessed 08 April 2013
6. Salesforce CRM. http://www.salesforce.com (2013). Accessed 08 April 2013
7. Jeffrey, K., Neidecker-Lutz, B. (eds.): The future of cloud computing—opportunities for European cloud computing beyond 2010. Expert group report, European Commission, DG INFSO (2010)
8. ten Hompel, M., Schmidt, T.: Warehouse Management: Automation and Organisation of Warehouse and Order Picking Systems. Springer, New York (2007)
9. Meinhardt, M.B., Rahn, J.: Empirical qualitative analysis of the current market situation in the context of cloud computing for logistics. In: ten Hompel, M., Rehof, J., Wolf, O. (eds.) Cloud Computing for Logistics, Lecture Notes in Logistics, Springer (2015)
10. Schuldt, A., Hribernik, K.A., Gehrke, J.D., Thoben, K.D., Herzog, O.: Cloud computing for autonomous control in logistics. In: Fähnrich, K.P., Franczyk, B. (eds.) 40th Annual Conference of the German Society for Computer Science (GI 2010), LNI 175, Gesellschaft für Informatik, pp. 305–310. Leipzig, Germany, 27 Sept-1 Oct (2010)
11. Kaisler, S., Money, W.H.: Service migration in a cloud architecture. In: 44th Hawaii International Conference on System Sciences (2011)
12. Goyal, P., Mikkilineni, R.: Policy-based event-driven services-oriented architecture for cloud services operation and management. In: IEEE International Conference on Cloud Computing (2011)
13. Scholz-Reiter, B., Rippel, D., Sowade, S.: Limitations in modeling autonomous logistic processes—challenges and solutions in business process modeling. In: IEEE International Symposium on Assembly and Manufacturing (ISAM), pp. 1–6, 25–27 (2011)
14. Bentounsi, M., Benbernou, S., Atallah, M.J.: Privacy-preserving business process outsourcing. In: IEEE 19th International Conference on Web Services (2012)
15. Fraunhofer Innovation Cluster "Cloud computing for logistics". http://www.ccl.fraunhofer.de/en.html (2013). Accessed 08 April 2013
16. Böhmer, M., Schmidt, M., Weißenberg, N.: Seamless interoperability in logistics: narrowing the business-it gap by logistics business objects. In: ten Hompel, M., Rehof, J., Wolf, O. (eds.) Cloud Computing for Logistics, Lecture Notes in Logistics, Springer (2015)
17. Holtkamp, B.: The logistics mall—an it-architecture for logistisc-as-a-product. In: ten Hompel, M., Rehof, J., Wolf, O. (eds.) Cloud Computing for Logistics, Lecture Notes in Logistics, Springer (2015)
18. Bochon, I., Ivens, V., Nagel, R.: Challenges of cloud business process management. In: ten Hompel, M., Rehof, J., Wolf, O. (eds.) Cloud Computing for Logistics, Lecture Notes in Logistics, Springer (2015)
19. Eggemann, J., Leveling, J., Weiß, N.: Business apps meet the challenge of covering continually changing logistics requirements. In: ten Hompel, M., Rehof, J., Wolf, O. (eds.) Cloud Computing for Logistics, Lecture Notes in Logistics, Springer (2015)

Empirical Qualitative Analysis of the Current Cloud Computing Market for Logistics

Maren-Bianca Wolf and Jonas Rahn

Abstract To determine the opinions, needs, and requirements of the market actors in the cloud computing market for logistics, Fraunhofer IML conducted an empirical market analysis with users and vendors of logistics and IT. Qualitative interviews were used to examine closely the potential of this concept and any barriers to it. The results of this analysis allow conclusions to be drawn about the future development in the logistics and IT industry.

1 Introduction

Cloud computing is now a reality. After the initial furore and cautious hesitation, the technology was able to hold its ground in the market and is now widely in use. Various sectors of the industry have recognized the advantage of distributed computing power and have put it to use in their respective areas of business. The numbers are clear: The rapidly increasing budgets of the past few years give us a sense of the size of the fundamental change in technology that it has brought with it —cloud computing has shaped and changed the IT market permanently in the same way that a disruptive innovation would.[1] Extremely volatile business areas with strongly fluctuating requirements such as logistics benefit greatly from the flexible distribution of resources in the cloud.

[1] Prüser [21].

M.-B. Wolf (✉) · J. Rahn
Fraunhofer Institute for Material Flow and Logistics, Dortmund, Germany
e-mail: Maren.Wolf@iml.fraunhofer.de

J. Rahn
e-mail: Jonas.Rahn@iml.fraunhofer.de

M. ten Hompel et al. (eds.), *Cloud Computing for Logistics*,
Lecture Notes in Logistics, DOI 10.1007/978-3-319-13404-8_3

2 The New Standard

In many places the use of cloud computing is now viewed as a matter of course and no further discussion is needed. For a few years this topic was still considered to be a future trend and although it was the cause of heated discussions it was almost never viewed as relevant for the policy of a company. This has clearly changed: According to the Federal Association for Information Technology, Telecommunications, and New Media (BITKOM); cloud computing has been selected as the hot topic for the information and telecommunications industry in 2013 for the fourth time.[2] The market research company Forrester Research predicted that for 2013 almost half of all North American and European companies will have a budget for investing in private clouds and many software development managers will plan to use cloud applications.[3]

3 Current Situation

According to estimates by the Experton Group, "German companies will put about 5 % of their IT costs into the cloud in 2013".[4] This means an increase in investment of circa 52 % over 2012: The costs for "cloud services, cloud integration, and consulting as well as cloud technology" in the business cloud will total approximately 4.6 billion euro in 2013. A growth of around 50 % has been forecasted for 2014. This is a total investment of 6.9 billion euro.[5]

According to other estimates, the cloud computing market in Germany will grow by 47 % in 2013 (business and consumer cloud together), like it did the previous year, and reach a volume of 7.8 billion euro (2012: 5.3 billion. euro). A sustained growth of between 30 and 40 % is also forecasted for the following year.[6] The cloud market in Germany is expected to reach a sales volume of approximately 20.1 billion euro in 2016. This is according to a report by the high-tech association BITKOM based on current forecasts by the Experton Group.

The European Commission released a statement announcing their new strategy for using cloud computing to boost the EU GDP annually by 160 billion euro by 2020.[7] According to BITKOM and the International Data Corporation (IDC), if the

[2] In 2010, 45 % of the surveyed ICT companies indicated that cloud computing was the most important IT trend. This was 62 % in 2011, 66 % in 2012, and 59 % in 2013. See the BITKOM press releases from 13 January 2010, 18 January 2011, 18 January 2012, and 16 Januar 2013.

[3] Staten [22].

[4] Experton Group 2013: Janata and Velten: Current market figures for the German cloud computing market—5 % hit barriers [13].

[5] Bayer [1].

[6] BITKOM press release from 6 March 2013 (current forecasts of the Experton Group): Sales in cloud computing rose to almost 8 billion euro [2].

[7] Europäische Kommission [5].

legal framework is modernized and the broadband infrastructure is expanded, then the annual boost to the GDP could be as much as 250 billion euro[8] and 3.8 million jobs could be created. Without political support the cloud computing boost to the EU GDP could be as low as 88 billion euro.[9]

4 Making the Shift to Cloud Computing

The number of companies making the shift to cloud computing has changed greatly in the past few years: Only 28 % of the companies surveyed in 2011 were open minded about cloud computing and interested in it. This number rose to 35 % in 2012. But also, surprisingly, the number of sceptics also increased from 38 % (2011) to 44 % (2012). The number of those who were undecided did go down, which indicates that companies improved their information policies and their uncertainty was dwindling. All of the discussions about the topic obviously helped decision-makers make up their minds: The number of those who were undecided shrunk from 33 to 20 %.[10]

The size of a company plays a role in how open they are to adapting the technology: 55 % of companies with more than 2,000 employees were open to it while smaller companies were clearly more reticent.[11]

5 Cloud Computing in Action

In 2011, 28 % of German companies were using at least one form of cloud computing (another 22 % were planning to implement it or discussing its implementation)—this number rose to 37 % in 2012 (planning rose to 29 %). Companies from the information and telecommunications and finance sectors were the most willing to adapt cloud computing. This group was followed closely by the transportation and logistics, chemical and pharmaceutical, automotive engineering, mechanical and plant engineering, and commerce sectors.

Some of the reasons for the positive attitudes of many companies towards cloud computing are lower IT administration costs, shorter implementation times for new applications and solutions, rapid scalability of IT services, improved mobility and geographically distributed access to IT resources, lower IT costs, increased organizational flexibility, improved performance and availability, and a higher capacity for innovation. The opportunity to tap into new customer groups and save costs

---

[8] Kempf [14].

[9] International Data Corporation (IDC): Quantitative Estimates of the Demand for Cloud Computing in Europe and the Likely Barriers to Up-take, p. 9 [12].

[10] KPMG AG, BITKOM & PAC: Cloud-Monitor 2013, p. 7 [17].

[11] Ibid., p. 8.

(hardware and personnel) also plays an important role. The main concerns are no longer with data protection but with the loss of data. Other risks that have been identified are a diminishing IT know-how and an unclear legal situation.[12]

6 Public Perception

Cloud computing still has a perception problem: According to a recent study from CSC, a global provider of IT services, 70 % of Germans are worried about their cloud data. They are primarily worried about their personal data in other countries and about the loss of control that comes with it. They also do not trust that the cloud service provider will perform the statutory delete requests reliably. The majority of the Germans who were surveyed (86 %) want proof of data protection and technical security from the cloud service provider. 82 % want the certification of the safety standards to be required by law.[13]

According to BITKOM and in spite of all of these concerns, every second Germany saves their photos, music, contacts, and calendars in the cloud.[14] When looking at only the 14–29 old, the average cloud usage rate goes up to 97 %.[15]

7 Public Clouds Versus Public Clouds

Eighty-three percent of users of private cloud services rated their experiences as positive in 2012, 25 % more than 2011. The rating for public cloud services dropped a bit in that same period: there were only 74 % satisfied users in 2012 instead of 81 %.

Thirty-four percent of Germany companies used private cloud solutions in 2012 (an increase of 7 % from the year before, 29 % were planning to implement one or discussing it) while 10 % decided to use a public cloud (an increase of 4 % from the year before, another 11 % were in the planning phase). Two thirds of the public cloud users were also using a private cloud. This means that many companies are using public cloud applications to supplement their private cloud. More than three quarters of all cloud users rated their experiences as positive.[16]

---

[12] Ibid., p. 4.

[13] CSC: IT-Sicherheit [4].

[14] Matz [19].

[15] Lüber [18].

[16] KPMG AG, BITKOM & PAC: Cloud-Monitor 2013, p. 4 [17].

8 Current Difficulties

8.1 Insufficient Utilization of Resources

The benefits of cloud computing are still not fully being utilized. One of the reasons for this is that there is still an inefficient usage rate of the cloud storage system: While the ideal usage rate of stationary data carriers is circa 50 %, only 17 % of the cloud storage is being used worldwide. The usage rate is a little bit higher in Germany at 26 %. However, if only small and medium-sized companies are considered then this falls to 7 %. Small and medium-sized companies are not using more than 90 % of their available—and paid for—cloud storage. It is no surprise that this is not helping to reduce costs.[17]

8.2 SLAs and Uniform Standards

Service Level Agreements (SLA) can help ensure that misunderstandings are avoided by specifying the scope and quality of the service in detail. However, both parties need the required know-how to create the SLA. The terms and conditions of the SLA should define the expectations between the two parties regarding availability of the services, the permissible failure rate, and the response time in case of failures. The SLA also has to define the consequences for non-adherence with the stipulated terms and conditions as well as the exclusions and limitations. It also has to contain the terms for terminating the SLA and the exact guidelines for returning and deleting the data (for example, the contract ends or one of the companies goes bankrupt).[18]

8.3 Data Protection and Safety

Although the cloud seems like it is ideal for the IT-intensive branches of the industry because of its scalability many of the players are showing restraint. Outstanding issues and unresolved preconceptions are still keeping some companies back from outsourcing applications to the cloud. In reality, many of the misgivings are unfounded: Cloud computing can actually help improve data protection, data availability, and performance.[19]

---

[17] Böckle [3].

[18] Kilz [15].

[19] KPMG AG, BITKOM & PAC: Cloud-Monitor 2012, p. 17 [16].

Even in the PricewaterhouseCoopers survey "Global State of Information Security Survey 2012" the majority of the companies surveyed (54 %) indicated that they improved their information security through the application of cloud services.[20]

9 Users of Cloud Computing

The number of cloud users has increased: In 2011, 28 % of companies were benefitting from the cloud while this number rose to 37 % in 2012. The number of companies planning to move to the cloud or at least discussing it climbed to 22 % in 2011 and to 29 % in 2012.[21] According to the IDC study "Cloud Computing in Germany 2012—Evolution of the Revolution", cloud computing was an official component of the IT strategy for 56 % of the companies surveyed and an official component of the business strategy for 34 % of the companies surveyed and this was by the middle of 2012.[22] The cloud seems to have established itself in medium-sized companies as well: According to the "IT Cloud Index Q4/2012", one quarter of all medium-size German companies were using Saas, laaS, and/or PaaS solutions by the end of 2012.

New trends such as big data, mobile applications, social business, and social collaboration encourage the use of cloud computing because they require global data availability as well as flexible computing and storage capacities.[23]

Small and medium-size companies benefit the most from the cloud: They are the ones who need custom cost—effective solutions but most companies do not have the budget to invest in numerous applications. With the advent of pay-per-use solutions, small and medium-sized companies now have access to solutions they could not afford before and they can now rent high-quality products without paying high licensing fees.

It really seems like the only way for small and medium-sized companies to stay competitive is for them to take the step into the cloud. Small companies just do not have the powerful and scalable IT infrastructures that they need to manage the constantly increasing volumes of data.[24]

[20] PricewaterhouseCoopers (PwC): Key findings from the Global State of Information Security Survey 2012, p. 24 [20].

[21] KPMG AG, BITKOM & PAC: Cloud-Monitor 2013, p. 10 [17].

[22] International Data Corporation (IDC): [11].

[23] Hofmann: Study conducted in Germany. IDC: » Die Verzahnung von Big Data und Cloud Computing wird enger « und ECIN: Cloud Computing 2013—Steigende Investitionen, Wachstumtreiber Social Collaboration und neue Sourcing Optionen [8].

[24] Techconsult GmbH; Hewlett-Packard GmbH [23].

10 Cloud Computing for Logistics

Several different surveys on cloud computing in the logistics sector have been conducted in the past few months and published as studies. One of them was an online survey conducted by the software provider INFORM GmbH which showed that 68.3 % of the surveyed companies are ready right now to use cloud computing for logistics tasks—only 12.7 % have actually done it. The reasons for this are a lack of familiarity with the topic (29.5 %) and the security concerns mentioned by almost half of the surveyed companies. The possibility of having to rely on an external service provider was a barrier to using cloud technology for 13 % of the surveyed companies. The lack of industry-specific solutions was an obstacle for another 5 %. There seems to be a wide range of reasons. Flexible access (38 %), reduction in operating costs (25 %), faster implementation times for business processes (18 %), platform independence (12 %), and access to IT resources that would not be possible without cloud computing (7 %) were identified as the benefits. According to the respondents, cloud computing solutions can be used for the communication between vendors and customers, controlling suppliers, and managing supply chain events.[25]

The current trend in the cloud is focussed almost exclusively on the use of a private cloud in the logistics industry. Only 6 % of the surveyed companies indicated in "Cloud Monitor 2013" that they use a public cloud solution. The logistics industry and the retail and wholesale industry are bringing up the rear in terms of public cloud usage—with the logistics industry slightly out in front.[26]

Even the findings of the study on the "Benefits of Cloud Computing for Logistics" that the Institute for Cloud Computing (IFCC) conducted on behalf of Axit AG confirmed the current trend: 60 % of the respondents were convinced that the future belongs to the cloud.[27] Logistics service providers are particularly open minded about the technology: almost 63 % of them are interested in the cloud (more than a third of them already have a cloud strategy). This is further proof that discussions about the cloud help to increase the trust in the fact that the cloud is here to stay—readiness grows with experience. As mentioned earlier, openness also increases with the size of the company: Companies with more than 1,000 employees show an above average interest in cloud computing.[28]

Small and medium-sized companies in particular often do not have enough IT expertise and capacity or the necessary capital to bridge the gap between the

---

[25] INFORM GmbH: Die Logistik und die Cloud—Vision oder Wirklichkeit? INFORM's trend barometer shows that despite security concerns it is medium-sized companies in particular who are open to the idea of using logistics IT services from the cloud [9].

[26] KPMG AG, BITKOM & PAC: Cloud Monitor 2013, p. 26 [17].

[27] Institute for Cloud Computing (IFCC); Axit AG: Studie zum » Nutzen von Cloud Computing für die Logistik « , p. 6 [10].

[28] Institute for Cloud Computing (IFCC); Axit AG: Studie zum » Nutzen von Cloud Computing für die Logistik « , p. 15 [10].

requirements and the status quo in logistics IT. Time is often the factor that prevents companies from developing flexible and innovative logistics solutions in a number of respects. The time it takes to implement the logistics systems of today is quite long compared to the length of time the system is used. As a result, business models and strategies in trade and commerce currently have a typical lifecycle of 2–3 years. An average implementation time of 6–10 months is shown when adequate (logistic) business processes are modelled. A new and more powerful IT solution is also required to provide the requested logistics services and this extends the implementation time by 12–24 months on average. The total usage time for the system (circa 3–5 years) is clearly too short for a company from a strategic investment point of view.

Pressure is also growing on the logistics industry because of some changes in legislation. In addition to the often cited toll charges, the new Working Hours Act and the necessity for electronic customs clearances both mean that new IT services will need to be used, for example for controlling driving times and rest periods. The same is true for the introduction of tracking and tracing for the continuous monitoring of goods.

It seems that the challenges placed on logistics services by both the industry and technology represent an exemplary area of application for Software as a Service solutions and cloud computing, which are viewed as the new paradigm for the provision of IT services and are considered to be a growth market.[29]

11 New Cloud Computing Concepts for Logistics: The Fraunhofer Innovation Cluster "Cloud Computing for Logistics"

The Fraunhofer Innovation Cluster "Cloud Computing for Logistics", which was started by the Fraunhofer Institute for Material Flow and Logistics IML and the Fraunhofer Institute for Software and Systems Engineering ISST, combines the design and structure of logistics and information technology services in one common design. The planned research and development work cumulated in the "Logistics Mall"as a central marketplace for everything from individual logistics IT functionality up to complete process chains, which can be offered as a product. This makes it possible for a company to put together the logistics services they need when they need them and operate them in the Logistics Mall. The opportunity to swap out an individual logistics IT service with the service offered by an alternative vendor at runtime does not only result in new business models but it also avoids dependencies on a service provider.

The Internet is both a technological platform and a signpost of the Logistics Mall and cloud computing is the future of the Internet. The main benefits of cloud

[29] Hamburger [7].

computing are going to be summarized here one more time. Cloud computing minimizes the costs and the barriers for starting the shift toward web portals. The reason for this is that the computing power is provided by the provider so no large investment in expensive infrastructure is needed. The scalability is another benefit because the computing power is adapted to the number of users. Cloud computing is the technology that is the safest bet for the future. One cloud consists of hundreds of thousands of servers that are replaced with newer models and always kept up to date. This means that the technology will never get old. Cloud computing also offers a high degree of availability, flexibility, and sustainability. Companies can reduce their ecological footprint because cloud computing helps them use their resources and systems more optimally and efficiently. The hottest topic is still the question of guaranteeing data protection in the cloud. Fraunhofer IML is paving a path here for the users and vendors in the Logistics Mall by offering a certification program for both vendors and products to ensure compatibility with the Logistics Mall.

The Logistics Mall uses Web 2.0 technology and was developed for the logistics of the future. With Web services as the basis, information and services within the Logistics Mall cannot only be requested but also new high value logistics process can be orchestrated. Like in an online store, the (virtual) user clicks on the desired services and thereby uses a global, open software architecture. The applications are hosted on the servers of the mall operators or the connected systems of the vendor. The customer only pays for what they use. The operator of the Logistics Mall handles the billing.

12 Results of the Empirical Analysis

Cloud computing will be indispensable to all industries in the future and the outsourcing of computer capacity to the cloud will be an essential component of every business model. The topic is still controversial today. Many market participants are worried about the security of cloud solutions and are not quite ready to make the jump.

The Fraunhofer Institute for Material Flow and Logistics IML examined the maturity of the logistics market for these solutions in a study. They asked users and vendors of logistics software about their attitude towards cloud computing technologies in a series of discussions.

Fraunhofer IML conducted the market analysis "Cloud Computing for Logistics" in 2011 and 2013 to identify and analyse the opinions, needs, and requirements of the users and vendors in the logistics industry in terms of both cloud computing and the Logistics Mall model. The project team conducted qualitative interviews and used the results of those interviews as the basis for examining the potential and barriers of this concept and determining which requirements needed to be implemented in the Logistics Mall. By examining the acceptance and readiness of those surveyed, they could draw conclusions about the attitudes of the logistics experts in the industry. To perform the analysis, they compared reference values from 2011 with the new values from 2013.

The study participants from the users group were from small and large companies from three branches of business: logistics service providers, industry, and wholesaler/retailer.

The study participants from the vendors group were from small and large providers of logistics solutions.

The results showed that the spirit of innovation of the users group is split right down the middle, which validated the theoretical assumptions that were made before the project began: It is still too early for every second participant to use a technology like the Logistics Mall. When asked more concretely 56 % of the participants from the users group would be willing to use logistics software in the cloud—if the conditions were right in their own company. This value has actually decreased by 8 % compared to the study conducted in 2011.

The interviews with the study participants from the vendors group in 2013 revealed that the innovative spirit of the vendors group is slightly less than it was in 2011. Half of the participants (49 %) are willing to sell their products or services through an external online store, which is decline of 9 % from 2011. A total of 67 % (70 % in 2011) would be willing to operate their applications in the cloud. In contrast to 2011, 10 % of the study participants in the vendors group are already in the concrete planning phase for using cloud architecture (Figs. 1, 2).

Can you imagine running your application in a cloud environment?

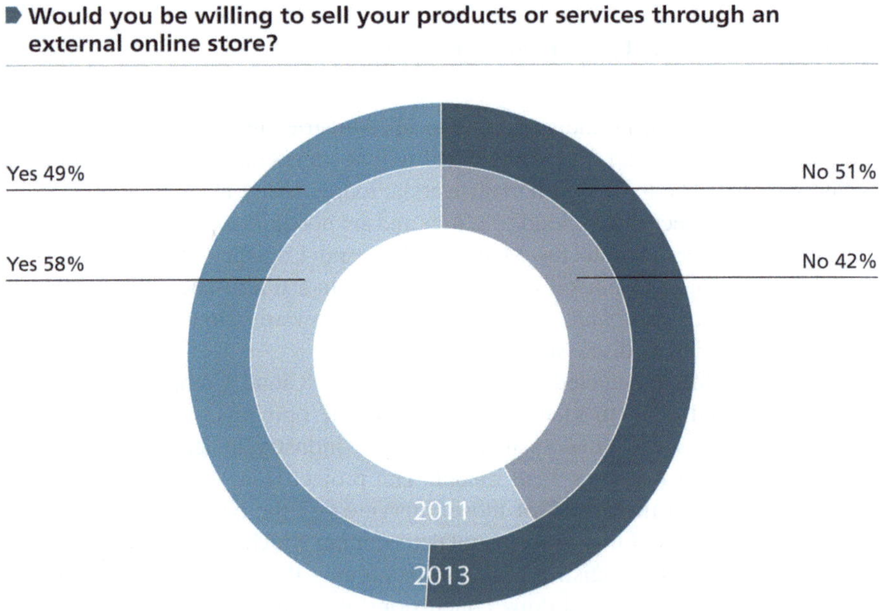

Fig. 1 The level of acceptance for selling products or services through an external online store (vendors)

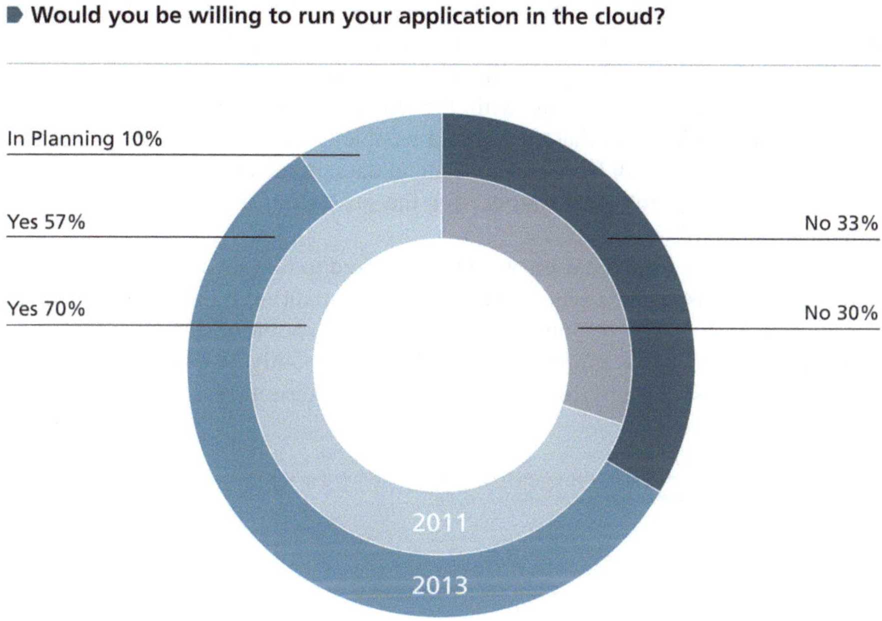

Fig. 2 The level of acceptance for operating applications in the cloud (vendors)

The agreement with these two fundamental Logistics Mall ideas diverges significantly with the size of the company. Small and medium-sized companies were more willing to change their views than larger companies. When the Logistics Mall was described in more detail, 58 % (75 %in 2011) of the vendors surveyed indicated that they would use it. Sixty-seven percent (70 % in 2011) of the vendors said that they were sure that their customers would be willing to use cloud computing in logistics. A big change in the answer to the question about the change in the relationship with the customer was seen when the 2013 data was compared to the 2011. Fifty-six percent (39 % in 2011) accepted the fact that as the vendor they were no longer the main contact point for the user but that the operator of the Logistics Mall held that role instead.

13 Opportunities and Risks

In addition to publicly known risks of performance and security (63 %), the users also named the following as areas of concern: failure of the cloud server (27 %), excessive complexity (23 %), problems communicating with existing systems (20 %), the insolvency of the vendor (14 %), the possible development of price dumping (7 %), the possible formation of a cartel (6 %), and a lack of differentiation from competitors (4 %).

According to the vendors, the main advantages for them and their customers are the opportunity to tap into new customer groups and the potential cost savings in personnel and hardware. They also hope that a synergy effect will result from the integration of their applications with the complementary applications of other vendors. This would mean that they would be able to offer customized solutions to their customers. They still have concerns about data protection and data security but these are more psychological barriers for the customers than problems with the technology (Fig. 3).

The participants from the users group were asked to rate several aspects of cloud computing in logistics on a scale from 1 (very important) to 6 (not important at all) as shown in Fig. 4. The respondents rated safe encryption of the data (1.5) and a guarantee that the applications run quickly and smoothly (1.6) as the two most important aspects. Twenty-four hour availability, customer support (2.2), and

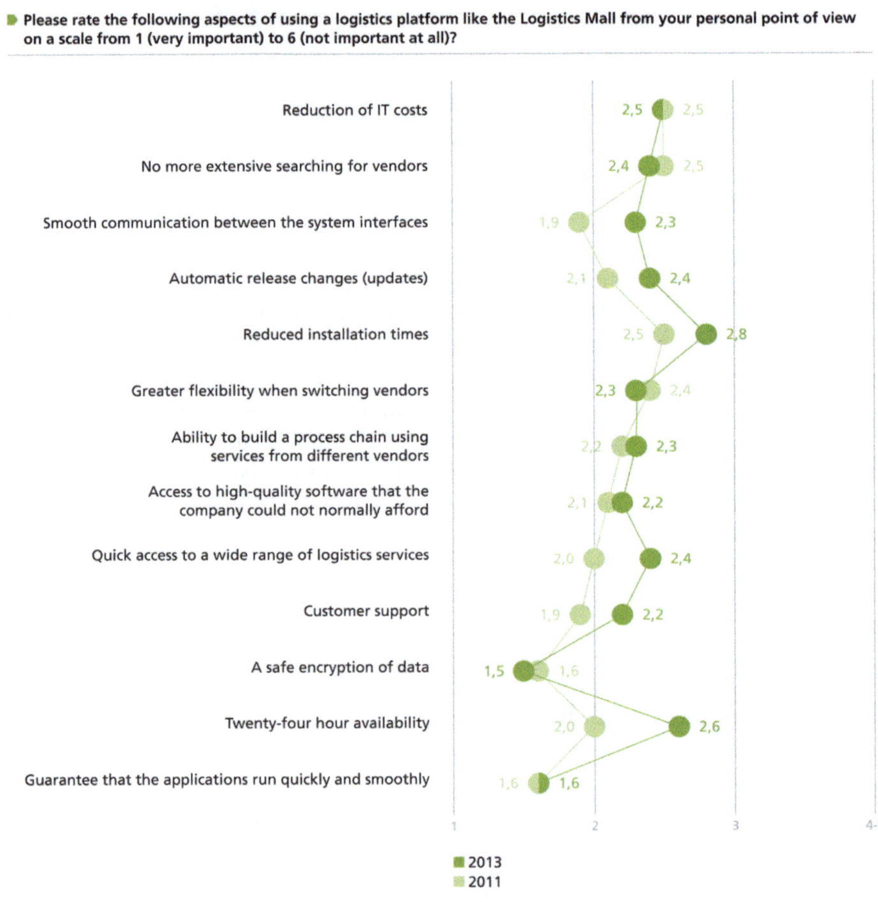

Fig. 3 Success factors for the user

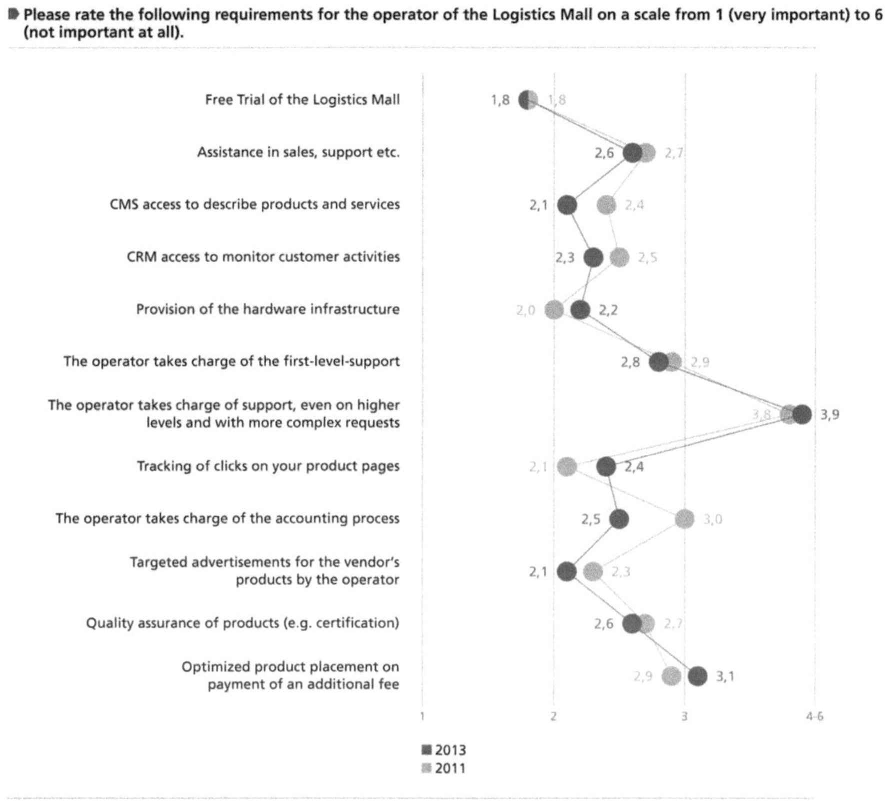

Fig. 4 Success factors for the vendor

access to high-quality software that the company could not normally afford (2.2) were rated as important. Smooth communication between the system interfaces and the ability to build a process chain using services from different vendors were both rated as important with a score of 2.3.

The main requirements of the vendor for the operator of the Logistics Mall are a free trial of the Logistics Mall (1.8), targeted advertisements for the vendor's products by the operator (2.1), vendor access to a Content Management System (CMS) to describe their own products and services (2.1), provision of the hardware infrastructure (2.2), and vendor access to a CRM system to monitor the activities of their own customers on the Logistics Mall website (2.3).

Twenty-six percent of users (22 % in 2011) are willing to enter into a contract with the operator of the Logistics Mall instead of the vendor. There is demand for the billing model (static or dynamic) and the contract periods and all possible configurations will soon be offered.

In 2011 37 % of the surveyed companies used WMS, whereas in 2013 only 32 % of the responding companies reported to operate such software. The utilization of TMS, however, has increased slightly: in 2011 30 % of the evaluated companies used this kind of software and in 2013 already 32 % of the companies did that could provide information about their software application. Regarding active ERP systems an even bigger discrepancy is identified: whereas in 2011 one third of the surveyed users confirmed the utilization, in 2013 already 46 % of the respondents applied an ERP system. The DMS use also increased from 27 % in 2011 to 34 % in 2013.

The reason that some participants in the users group have never used logistics software is that their company is too small to have special systems to support their logistics processes and, as a result, they did not have the budget for it. Twenty-five percent (30 % in 2011) of those who have used logistics software indicated that they have thought about changing vendors. The willingness to change is low because they think that the costs associated with making this change are too high and cannot even be justified by the improvements in performance and functionality they would get by changing vendors. Often they can get their current vendor to make some improvements by complaining so it seems to them that it is no longer necessary to make a change.

14 Results

The results clearly show the spirit of innovation of German managers: with an acceptance rate of 56 % this value has sunk by 12 % compared to 2011. Despite this decline, this number still shows that more than half of the respondents are ready and willing to rent logistics software from the cloud today. They answered the question "Can you imagine using logistics software that is not running locally on computers at your company but instead on servers in the Internet?" with yes. Those who answered no revealed in the qualitative interviews that one of the reasons was that they did not feel familiar enough with the topic. It is important to note that the concept of the Logistics Mall had not been explained to them yet when this question was asked. The goal of this question was to determine their attitude towards the idea of cloud computing in logistics. Some of those who said no to cloud computing in the survey explained that it was because they thought the technology was not developed enough for them to make a shift in their IT strategy.

15 Conclusion and Outlook

The increasingly homogenous range of services and products offered in the logistics market are making it harder and harder for a customer to select the right vendor. From the vendor's perspective (for small and medium-sized companies), it can be difficult to stay competitive because the main focus is always on the price and certain well-known companies have put their stamp on the market over the years. Large companies also face the challenge of staying competitive because customers keep asking for more and more features. Customers want flexible and tailor-made solutions and they want them "on demand". This demand from customers has driven the need in the future for a suitable marketplace where companies can offer their logistics solutions. Just like a shopping mall, a B2B customer wants to be able to select the products and services they want and combine them with others. To be able to continue to increase customer satisfaction and also satisfy all demands, two disciplines need to mesh and become better integrated in the future—logistics and marketing.

> Customers do not look for goods or services per se, they look for solutions.[30]

The quote clearly shows that customers are thinking bigger than just services and products. They want a solution for their problem. They want added value at the end of a process. It all comes back to individuality. The Logistics Mall as a solution for providing custom service offerings fosters the interactive character of the solution and leads to increased customer satisfaction. This new approach, one of viewing and using the Logistics Mall not just as a platform for products and services but as a solutions platform, needs to be lived. In the future this will mean for the customer that none of their problems will go unsolved.

The results of the market analysis make it clear that the time for cloud computing and, in particular, cloud computing in logistics is here. The logistics market is already moving in that direction.

The business culture will change over time. The focus will be on custom solutions. The network concept that the Logistics Mall represents leaves a lot of leeway. Each company—no matter how big or small—has the chance to be a part of this development and help shape it: The experts are already saying that the Logistics Mall will become the largest logistics platform in the industry.

References

1. Bayer, M.: Die Cloud wird gesellschaftsfähig, in: COMPUTERWOCHE, online, 6 Februar 2013: http://www.computerwoche.de/a/die-cloud-wird-gesellschaftsfaehig,2532192
2. BITKOM Presseinformation vom 6. März 2013 (aktuelle Prognosen der Experton Group): Umsatz mit Cloud Computing steigt auf fast 8 Milliarden Euro, online: http://www.bitkom.org/de/markt_statistik/64086_75301.aspx
3. Böckle, R.: Symantec-Studie. Versteckte Kosten in der Cloud, in: ChannelPartner, online, 5. Februar 2013: http://www.channelpartner.de/2601143
4. CSC: IT-Sicherheit: 70 % der Deutschen haben Angst um ihre Cloud-Daten, online, 6. Dezember 2012: http://www.csc.com/de/press_releases/93261-it_sicherheit_70_%25_der_deutschen_haben_angst_um_ihre_cloud_daten
5. Europäische Kommission: Mehr Schwung für das Cloud Computing, online, 27. September 2012: http://ec.europa.eu/news/science/120927_de.htm

[30] Grönroos: Service Management and Marketing. A Customer Relationship Management Approach, p. 4 [6].

6. Grönroos, C.: Service Management and Marketing: A Customer Relationship Management Approach, Wiley, Chichester (2000)
7. Hamburger Abendblatt: Microsoft und Telekom erwarten Durchbruch für Cloud Computing, online, 07.10.2010:http://www.abendblatt.de/wirtschaft/article1654862/Microsoft-und-Telekom-erwarten-Durchbruch-fuer-Cloud-Computing.html
8. Hofmann, K.: In Deutschland durchgeführte Studie. IDC: » Die Verzahnung von Big Data und Cloud Computing wird enger « , in: IT-Business, online, 1. Februar 2013: http://www.it-business.de/index.cfm?pid=4840&pk=393096&print=tr
9. INFORM GmbH: Die Logistik und die Cloud—Vision oder Wirklichkeit? INFORM-Trendbarometer zeigt: Trotz Sicherheitsbedenken ist vor allem der Mittelstand offen für Logistik-IT-Services aus der Cloud, online, 18. Januar 2013: http://www.inform-software.de/produkte/invent-xpert/news/article/die-logistik-und-die-cloud-vision-oder-wirklichkeit/
10. Institut für Cloud Computing (IFCC); Axit AG: Studie zum » Nutzen von Cloud Computing für die Logistik « , online, Juli 2012: http://www.axit.de/cloudlogistik/
11. International Data Corporation (IDC): Cloud Computing in Deutschland 2012—Deployment-Modelle und Management, Integration, Security und Compliance im Fokus, online, Juli 2012: http://www.idc.de/consulting/mc_cloud2012_reg.jsp
12. International Data Corporation (IDC): Quantitative Estimates of the Demand for Cloud Computing in Europe and the Likely Barriers to Up-take, online, Juli 2012: http://ec.europa.eu/information_society/activities/cloudcomputing/docs/quantitative_estimates.pdf
13. Janata, S., und Velten, C. (Experton Group): Aktuelle Marktzahlen zum deutschen Cloud-Computing-Markt—5-%-Hürde erreicht, online, 30. Januar 2013: http://www.experton-group.de/research/ict-news-dach/news/article/aktuelle-marktzahlen-zum-deutschen-cloud-computing-markt-5-%25-huerde-erreicht.html
14. Kempf, D.: Vortrag im Rahmen der Pressekonferenz zum Cloud Monitor 2013, online, 21. Februar 2013: http://www.bitkom.org/files/documents/BITKOM_Vortrag_PK_Cloud_Monitor_21_02_2013_final_web.pdf
15. Kilz, S.: Wasserdichte Verträge und SLAs aushandeln, in: Computerwoche, online, 27. Februar 2013: http://www.computerwoche.de/a/wasserdichte-vertraege-und-slas-aushandeln,2533212
16. KPMG AG, BITKOM & PAC: Cloud-Monitor 2012, Hannover, online, 6. März 2012: http://www.kpmg.de/docs/cloud-monitor-20120529.pdf
17. KPMG AG, BITKOM & PAC: Cloud-Monitor 2013, Hannover, online, 6. März 2013: http://www.bitkom.org/files/documents/BITKOM_KPMG_PAC_Studie_Cloud_Monitor_2013.pdf
18. Lüber, K.: Die Cloud—des einen Freund, des anderen Feind, in: Die Welt, online, 27. Februar 2013: http://www.welt.de/sonderthemen/cebit-2013/article113928252/Die-Cloud-des-einen-Freund-des-anderen-Feind.html
19. Matz, A.: Die Wolke im Griff, in: Hamburger Abendblatt, online, 5. April 2013: http://www.abendblatt.de/ratgeber/article115026738/Die-Wolke-im-Griff.html
20. PricewaterhouseCoopers (PwC): Key findings from The Global State of Information Security Survey 2012, online, September 2011: http://www.pwchk.com/webmedia/doc/634653330562192188_rcs_info_security_2012.pdf
21. Prüser, S.: Die Cloud—In aller Munde, aber noch lange nicht im Handel, auf: heise resale, online, 16. November 2011: http://www.heise.de/resale/artikel/Die-Cloud-In-aller-Munde-aber-noch-lange-nicht-im-Handel-1331924.html
22. Staten, J.: 2013 Cloud Predictions: We'll finally get real about Cloud, online, 3. Dezember 2012: http://blogs.forrester.com/james_staten/12-12-03-2013_cloud_predictions_well_finally_get_real_about_cloud
23. Techconsult GmbH; Hewlett-Packard GmbH: IT-Cloud-Index, Q4/2012, online, April 2013: http://www.techconsult.de/index.php/43-uncategorised/231-clouindex1-13

The Logistics Mall—An IT-Architecture for Logistics-as-a-Product

Bernhard Holtkamp

Abstract The Logistics Mall extends the cloud service model from Software-as-a-Service (SaaS) to Business Process-as-a-Service (BPaaS). Its architecture, hence, spans from a web-based public B2B marketplace for simple apps, traditional applications and even logistics process templates to customer-specific cloud based execution environments for instantiated process models and therein included applications. For communication with the mall outside world a specific gateway is provided that controls inbound and outbound message flow.

Keywords Logistics mall · Cloud computing · Software-as-a-service (SaaS) · Business process-as-a-service (BPaaS)

1 Introduction

Currently the time for implementing new logistics systems is long compared with their usage time. In trade and industry business models and strategies have a life cycle of 2–3 years. The development of new logistics processes often takes 6–10 months. If new IT systems are needed to support these processes, implementation time increases to one or even up to 2 years. This is too long and too expensive for strategic investments. The use of logistics specific cloud services is therefore considered as a viable solution.

Hence, the Logistics Mall provides virtualization of IT resources and business processes to relieve especially small and mid-size production and logistics companies from investments in the development of individual software solutions including expensive software licenses and from operating costs. Thus, these companies can focus on their core business.

B. Holtkamp (✉)
Fraunhofer Institute for Software and Systems Engineering ISST,
Emil-Figge-Str. 91, 44227 Dortmund, Germany
e-mail: Bernhard.Holtkamp@isst.fraunhofer.de

M. ten Hompel et al. (eds.), *Cloud Computing for Logistics*,
Lecture Notes in Logistics, DOI 10.1007/978-3-319-13404-8_4

Logata is a key partner of the Fraunhofer innovation cluster "Cloud Computing for Logistics" that develops the Logistics Mall. Logata is the provider and operator of the Logistics Mall as a public cloud solution. Low margins and growing competition in the logistics market on the one hand and Logata's position as an SME on the other hand require a low cost solution. To avoid expensive license fees for software components the Logistics Mall is implemented as an aggregation of open source software solutions.

In the following we have a closer look at the Logistics Mall architecture. Following the TOGAF Architecture Development Model [1] we start with the Business Architecture. The architecture vision of the Logistics Mall is already described in detail in [2]. In a next step we discuss some design decisions that are relevant for the understanding of the technical architecture that is presented in subsequent sections of this document. We present the Logistics Mall version one that applies the Software-as-a-Service paradigm first and continue with enhancements of this version towards a higher level of integration of applications within the Logistics Mall and with external applications up to the level of Business Process-as-a-Service in the final version of the Logistics Mall. We close with a look to the current status of usage and with an outlook to further developments.

2 Business Architecture

The description of the business architecture focuses on three aspects: business strategy, organization and key business processes to be supported by the Logistics Mall.

2.1 Business Strategy

Logata is the Logistics Mall provider and thus the "face to the customer". Logata aims at providing a broad high-quality offering. This means, like in a shopping mall, a balanced portfolio of products and competing suppliers. To guarantee high quality suppliers and their products are certified. The focus is on suppliers of logistics IT applications and of business process support. Hence, the Logistics Mall provides a rich portfolio of services for suppliers to appear as an attractive sales and distribution channel. That includes CRM functionality for offerings and follow-ups of such offers, Infrastructure-as-a-Service, help desk and 1st level support and last but not least billing support for suppliers towards their customers.

In addition to the role as Logistics Mall provider Logata also intends to act as a supplier to offer their own solutions in the mall. The target group is formed by SMEs from the logistics and production domains with annual revenues in the range of 70 Mio. Euros or an IT budget of about 1.4 Mio. Euros. As pricing model a monthly flat rate is envisaged. The goal is to provide customers with IT solutions that reduce their IT costs by 30 %.

In a later stage the target group will be extended towards global players.

2.2 *Organization*

Logata is the provider of the Logistics Mall as a public cloud solution. In that role Logata is in charge of acquiring suppliers of logistics IT applications and to perform marketing and PR activities to create awareness for the Logistics Mall in the customer target groups.

As the "face to the customer" Logata also provides a helpdesk and 1st level support to end users, i.e. registered customers that have booked services in the Logistics Mall. For suppliers Logata acts as a clearinghouse, doing billing and invoicing towards the suppliers' customers in the mall and billing towards the suppliers for the use of Logistics Mall services.

Suppliers offer IT applications as "Software-as-a-Service" solutions and provide 2nd level support for their applications.

In the first phase, Logata is also the technical operator of the Logistics Mall.

2.3 *Business Processes*

When looking at business processes to be supported by the Logistics Mall we have to differ between the following stakeholder roles: Logistics Mall provider, Logistics Mall operator, potential mall customer, logistics IT supplier and end user. On the end user side we can further differentiate between administrator, normal user and manager as a privileged user.

Potential customers of the Logistics Mall can navigate through the public part of the Logistics Mall homepage on the Web and can register for becoming a mall customer.

Registered end users can book applications, services and processes in the Logistics Mall. They can use the booked functions, receive usage reports and billing information. In case of problems the users can contact the helpdesk or can submit trouble tickets.

Registered suppliers can apply for the inclusion of their applications and services in the Logistics Mall portfolio. If a product is accepted the supplier gets the means to manage the product in the Logistics Mall catalogue. Suppliers also receive usage reports and billing information. They have to provide 2nd level support for their products. In case of problems they can also contact the helpdesk or submit trouble tickets.

The Logistics Mall provider is in charge of marketing and PR processes to attract suppliers and users for the offered solutions. The Logistics Mall provider checks registrations and endorses or rejects applications. The provider also performs the administration of the Logistics Mall, including customer management, billing and reporting. Logistics Mall customers are suppliers of IT applications as well as users who book such services and use them in the Logistics Mall cloud. The Logistics Mall provider also runs the helpdesk and provides 1st level support to its customers.

The Logistics Mall operator runs the Logistics Mall infrastructure including monitoring and backup of the Logistics Mall. The operator also sets up usage environments for new customers and integrates new applications in the mall environment.

3 Design and Implementation Decisions

Security and maintenance aspects were driving forces of the following design and implementation decisions. Security is a big issue anyway regarding the acceptance of cloud computing, especially in Germany. Motivating companies to use business critical applications in the cloud implies the adoption of adequate security measures. Providing a company exclusive environment is an adequate measure.

Vendor lock-in is another issue that prevents companies from migrating their applications into the cloud of a particular cloud service provider. An exclusive environment per customer makes it easier for the cloud service provider to return customer data to their customer and to remove the customer from the cloud without leaving security and privacy relevant traces.

3.1 Design Decisions

A key decision for the Logistics Mall is the separation of the commercial part from the usage part. That leads to a Logistics Mall Marketplace (MMP) with the portfolio of offered applications, services and processes. Customers can register here and registered customers get access to their private part of the marketplace to book mall products, submit calls for tenders to suppliers or read their usage reports or billing information. Suppliers get services to manage their products in the mall catalogue, i.e. edit product descriptions or their company profile.

Users, having booked applications in the mall, get access to their services via a Customer Access Framework (CAF). Each customer company gets its own CAF to ease the management of service level agreements as well as backup/restore and the management of maintenance windows.

To reduce central administration efforts, self-administration of customers is applied to user management, to communication management via the Logistics Mall Gateway and to peripheral devices at customer premises that are used by applications in the mall.

A single sign on service is provided that covers the MMP, a customer CAF and the applications integrated in the CAF.

3.2 Implementation Decisions

As mentioned before a general implementation decision with regard to license fees was the use of Open Source software as far as possible. An exemption was made for the virtualization layer. Regarding the experience with VMware and the product quality at the time the decision was made it was decided to use VMware vSphere for the virtualization layer.

After the evaluation of various portal software solutions [3] Liferay was taken as a basis for portal development. For security and maintenance reasons it was decided to have different portal instances for MMP and CAF. Moreover, each customer company gets its own CAF portal on its own virtual machine.

Components that are shared between MMP and CAFs (e.g. user management or basic services (reporting, trouble ticketing, and accounting)) are implemented in a Logistics Mall Infrastructure layer (LMI).

As long as applications are not multi-tenant ready each customer gets their booked application instance on its own virtual machine.

4 Logistics Mall (Stage 1)

To enable a fast time-to-market the Logistics Mall is developed in steps. In the first step the Logistics mall offers logistics IT applications applying the "Software-as-a-service" paradigm (Fig. 1).

Customers book one or more applications in the Logistics Mall Marketplace (MMP) and get access to the booked applications through their Customer Access Framework (CAF). In the following we have a brief look at the logical architecture

Fig. 1 Logistics Mall as public cloud SaaS platform

of the Logistics Mall, followed by a closer look at the technical architecture. Finally, MMP and CAF are discussed in more detail.

4.1 Logical Architecture

The logical architecture of the Logistics Mall can be seen from different perspectives. From the viewpoint of the stakeholders the Logistics Mall looks as depicted in the following figure. A purchaser of a logistics service provider or of a production company basically sees the MMP. He uses the MMP for collecting information about the Logistics Mall portfolio for registration as a customer. After registration he uses the MMP for booking applications. The management of the customer company might also use the MMP for accessing usage reports and billing information. Customer administrators use the customer exclusive CAF for managing users, their roles and access rights, and for managing peripheral devices that must be available to booked applications. Users of the customer company log into their CAF to access the booked applications.

In case of problems with the Logistics Mall or with booked applications any user can submit trouble tickets via MMP or CAF.

The 1st level support of the Logistics Mall provider uses the Web interface of the integrated trouble ticketing system to handle trouble tickets from their customers (Fig. 2).

From a technical perspective the Logistics Mall looks as depicted below. MMP and CAF portals reside on top of the Logistics Mall Infrastructure (LMI). LMI and the portals use components from the basic software platform underneath.

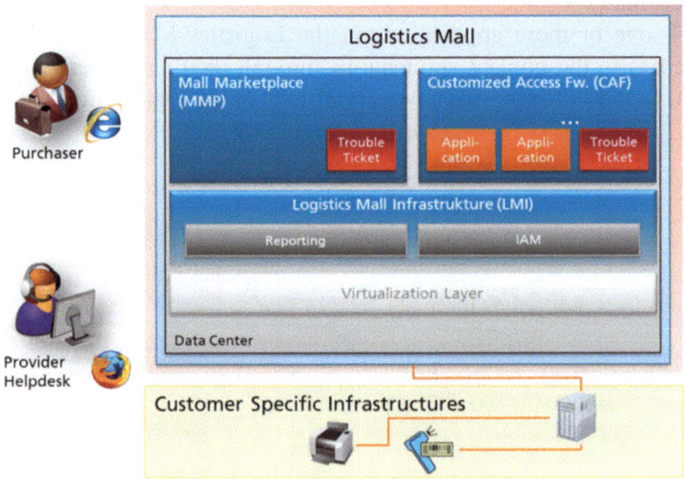

Fig. 2 Logistics Mall (stage 1) stakeholder view

Fig. 3 Logistics Mall logical architecture from a technical viewpoint

The bottom layer provides the virtual machines for the upper layers. It is implemented by VMware vSphere 4 Essentials Plus (Fig. 3).

4.2 Technical Architecture

On a gross level MMP and CAFs have the same architecture. The frontend part consists of the presentation layer and a service integration component. The backend part is composed of the service layer and the persistence layer. The components are discussed in more detail below (Fig. 4).

Figure 5 illustrates the implementation of CAF components using basic open source software platform components. My WMS and Puzzle represent applications that are integrated in the CAF. The black arrows show the control flow from a user initiated request in the browser down to persistence layer components.

4.3 Presentation Layer

The Logistics Mall presentation layer provides different interfaces for the different user roles identified in Sect. 2.3 as well as interfaces to functionality independent of a user's role. Furthermore, interfaces are different for MMP and CAFs.

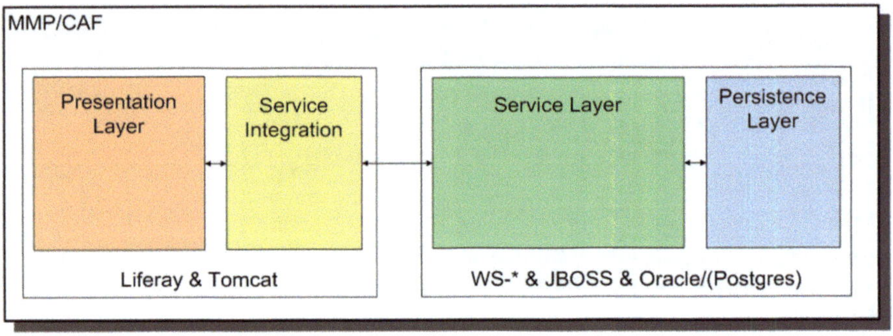

Fig. 4 Logistics Mall gross architecture

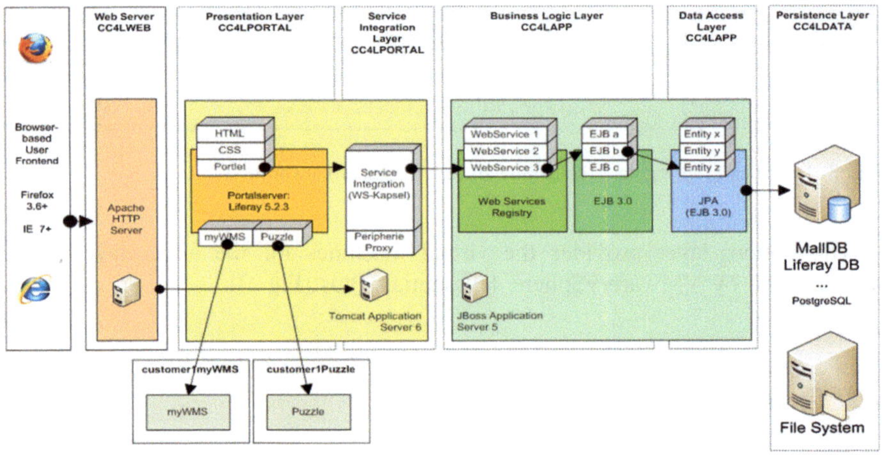

Fig. 5 Logistics Mall (stage 1) technical architecture

The MMP presentation layer provides stakeholder specific interfaces. An unregistered visitor gets access to public information like e.g. application portfolio or supplier catalogue and to the registration service. When a buyer (from a registered customer company) logs in, the MMP provides access to the list of booked products, to the related list of suppliers and to the list of pending orders. When a supplier logs in, the MMP provides service interfaces to product management, offers management, customer contact management, product registration and supplier activities. The mall operator is provided with content management functionality for news and help, customer registration management services, system administration services, services related to operator activities and with product management services (Fig. 6).

In addition to stakeholder specific services the MMP provides common services for single sign on, customer and user administration, reporting and reporting

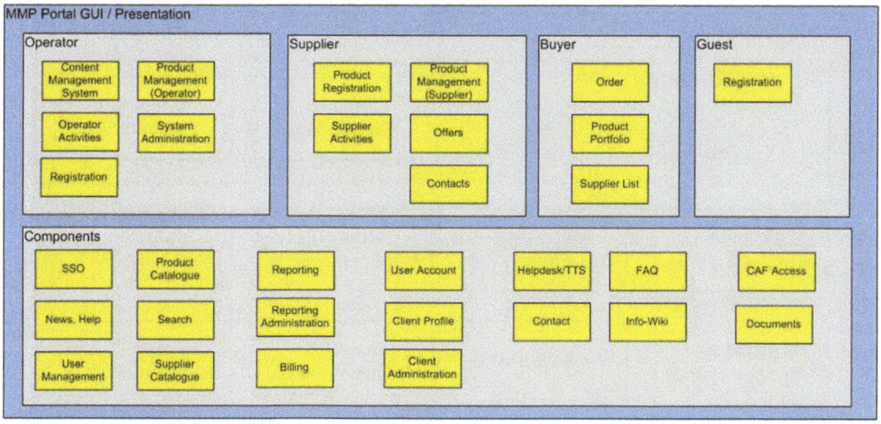

Fig. 6 MMP presentation layer components

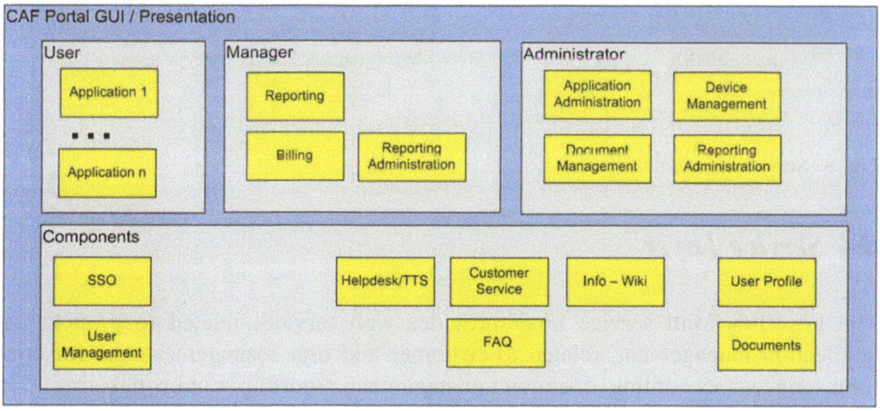

Fig. 7 CAF presentation layer components

administration, accounting, news access, trouble ticketing, searching, document management and CAF access.

The CAF presentation layer provides service interfaces for the user roles of a customer. A regular user gets access to the applications he has a role in. Managers get access to reporting and accounting services and to reporting administration. Administrators are provided with service interfaces for application management, document management, reporting management, and for the management of peripheral devices. All customer user roles get interfaces to common services like single sign on, user administration, helpdesk and document management (Fig. 7).

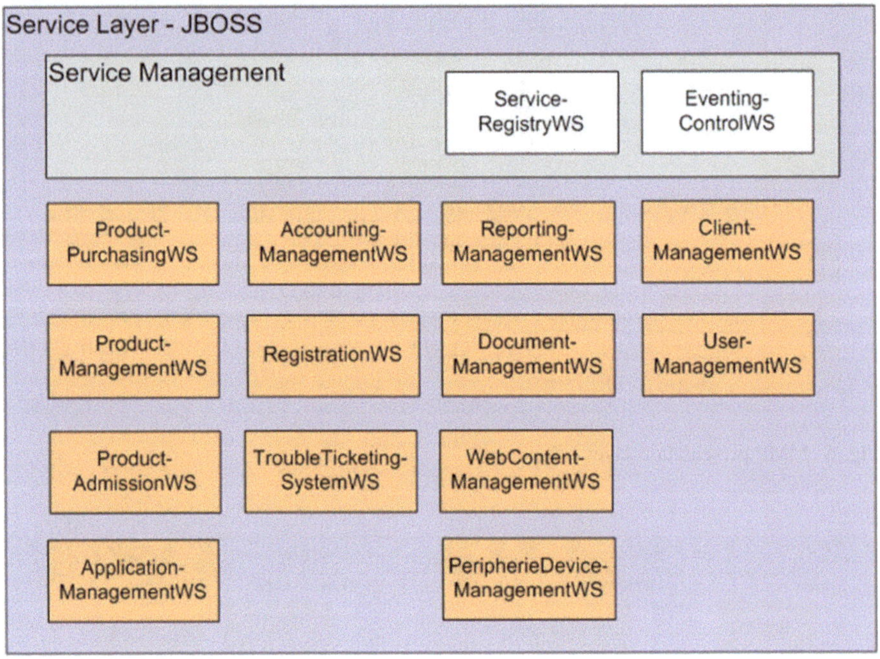

Fig. 8 Service layer components

4.4 Service Layer

The Logistics Mall service layer provides web services related to product and application management, related to customer and user management and for common services (accounting, document management, reporting, trouble ticketing). The Web Content Management services support MMP content management. The Peripheral Device Management services support the management of peripheral devices on a customer's premise that are connected to a booked application in the Logistics Mall cloud (Fig. 8).

4.5 Persistence Layer

The persistence layer manages the Logistics Mall non-volatile data. Key components are the product catalog, tenant registry, user registry, roles registry and MMP content. The product catalog is divided into two parts: the product records contain information about the product types offered in the MMP, the product instance

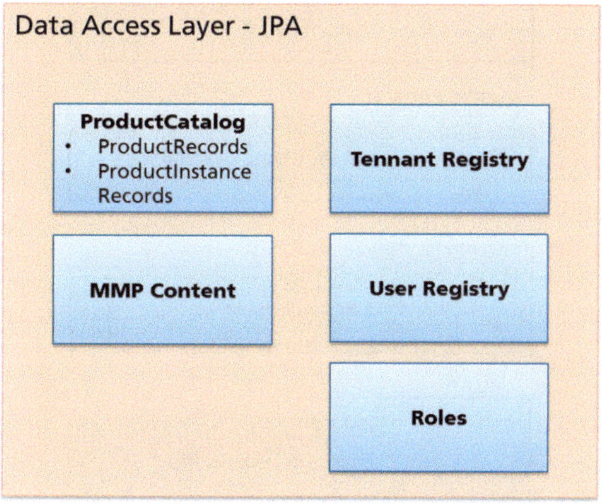

Fig. 9 Persistence layer components

records contain information about the instances of product types that are set up when booked by a customer. The tenant registry manages company related information of suppliers and end user companies that have registered. The roles registry keeps the roles that are supported by the products in the MMP product catalog. The user registry keeps the user account information of the Logistics Mall tenants (Fig. 9).

4.6 Implementation

The Logistics Mall is implemented by means of open source software. Key components are already mentioned in Sect. 4.2 when describing the technical architecture. Figure 10 shows the implementation layers with their related technologies.

The Logistics Mall components are deployed on different virtual machines as illustrated by Fig. 11.

In our development and test environment the Logistics Mall virtual machines run on the physical infrastructure depicted in Fig. 12.

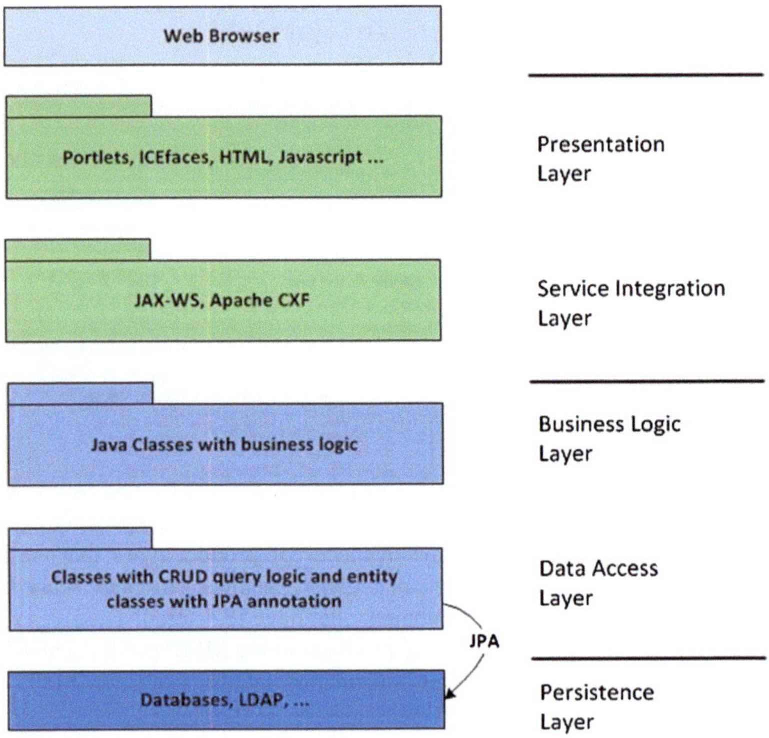

Fig. 10 Logistics Mall implementation layers

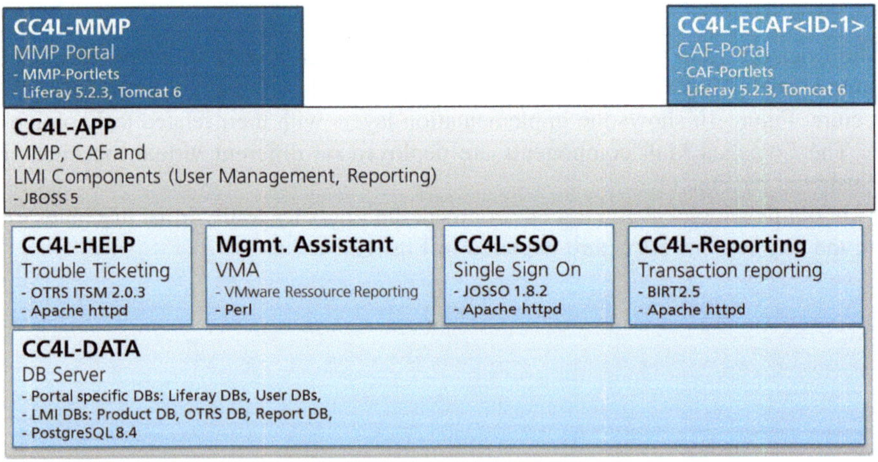

Fig. 11 Logistics Mall deployment schema

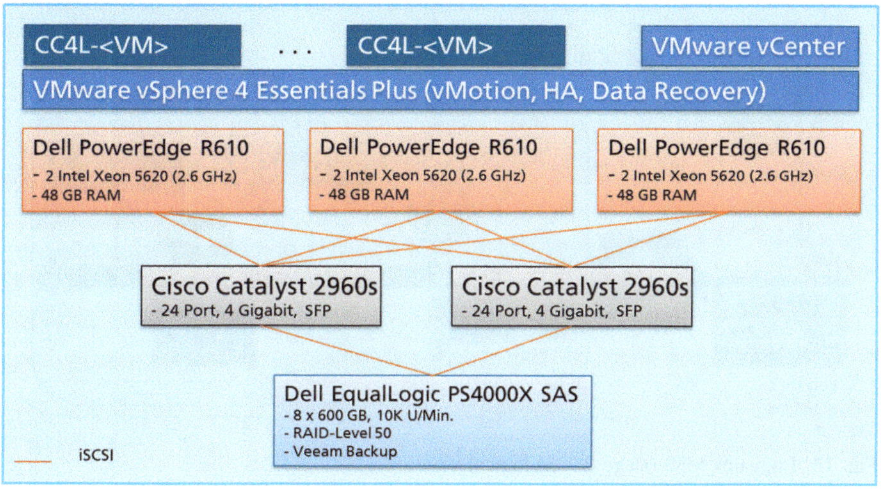

Fig. 12 Logistics Mall physical infrastructure (development and test environment)

5 Logistics Mall (Stage 2 + 3)

The Logistics Mall in stage 2 extends the SaaS model supported in stage 1 by providing application integration support for applications in a CAF as well as for applications in the mall with applications outside. Integration support in a CAF is provided by the Logistics Mall Bus and by the Business Object Repository. Integration of CAF internal applications with applications outside of the Logistics Mall is supported by means of the Logistics Mall Gateway. The Logistics Mall in stage 2 and 3 supports not only traditional applications but apps as well. From the Logistics Mall point of view an app executes IT-based activities as part of logistics processes. Apps can be combined to complex workflows. They can be interactive or fully automated. The functionality ranges between simple services in the sense of SOA and complex applications. Apps communicate by exchanging business objects via the Logistics Mall Bus within a CAF or via the Logistics Mall Gateway with external parties. They can also share objects within the Business Object Repository. More details on Logistics Mall Apps are given in [4].

The Logistics Mall Bus enables an asynchronous exchange of Logistics Mall Business Objects between connected applications within a CAF. The bus provides a web service interface that accepts a business object from a sending application. The bus buffers a business object and passes it to a receiving application via web service request.

The Business Object Repository stores instances of Logistics Mall Business Object types. The repository can store the objects that are used within an application but can also act as a shared data source for multiple applications. A detailed discussion of Logistics Mall Business Objects and their repository can be found in [5] (Fig. 13).

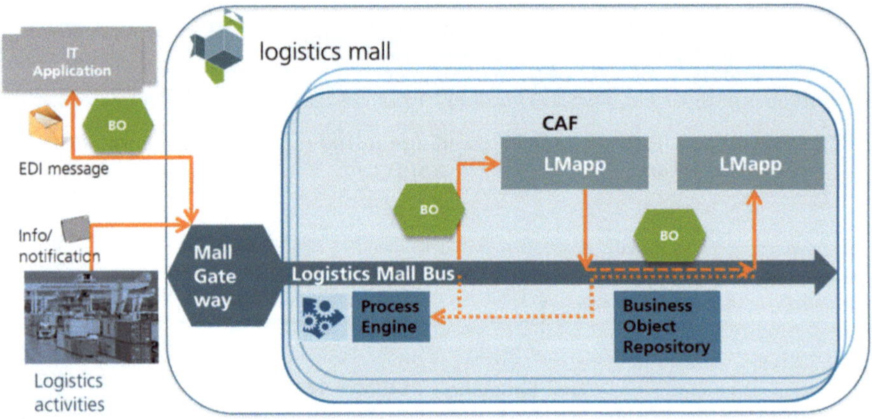

Fig. 13 Logistics Mall (stage 2 + 3) logical view

The Logistics Mall in stage 3 also provides for logistics business process management from process design to process execution. The web-based Logistics Process Designer supports Logistics Mall compliant design of logistics processes by combining apps from the mall to executable workflows. These processes can then be deployed on a process engine that forms an integral part of a CAF in stage 3. For more details on process management in the Logistics Mall we refer to [6].

5.1 Logistics Mall Bus

The Logistics Mall Bus represents the internal communication infrastructure of a CAF. It provides web service interfaces for the exchange of Logistics Mall Business Objects between connected applications. Business objects are transferred as Business Object Documents according to OAGIS [7] in XML format.

The Logistics Mall Bus provides the TransferBOD webservice server that is called by an application or by to pass a business object to the bus. The bus transfers a business object into the queue of the receiving application. For each connected application the bus maintains such a queue for buffering BODs until they can be passed to the receiving application. The ReceiveBOD webservice client transfers the BODs from an application's queue via webservice call to the application connector (see Fig. 14). The bus log records all transfers so that the flow of business object can be traced if necessary.

Figure 15 shows the Logistics Mall in stage 2 from a technical perspective.

The Logistics Mall Bus is implemented by means of the open source talend enterprise service bus [8]. For the buffering of messages Apache JMS queues are used [9].

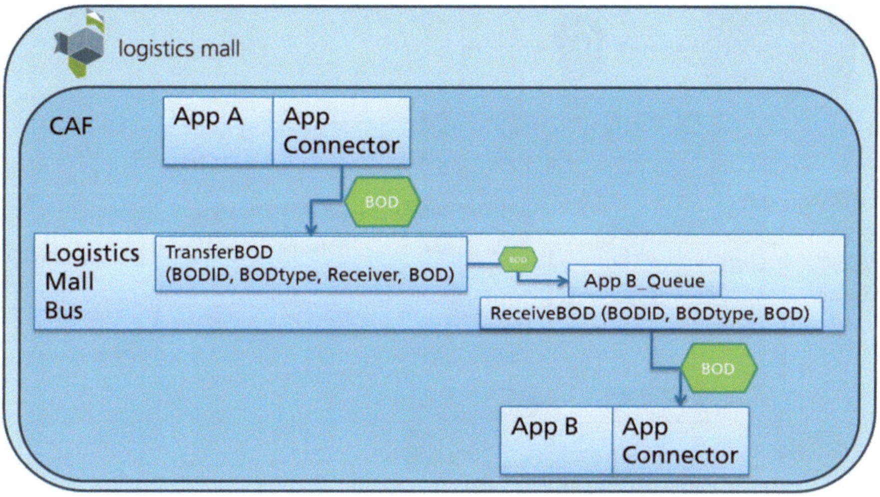

Fig. 14 Transfer of BODs between applications in a CAF

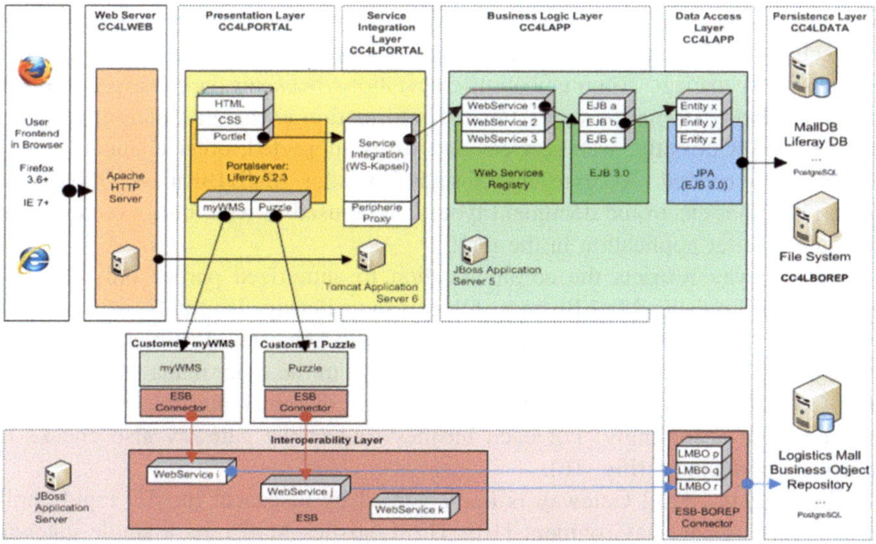

Fig. 15 Logistics Mall (stage 2) technical view

5.2 *Logistics Mall Gateway*

The Logistics Mall Gateway is the interface of the Logistics Mall to the outside world. The gateway coordinates and controls the communication between applications in the mall and their outside peers. Information exchange is only allowed

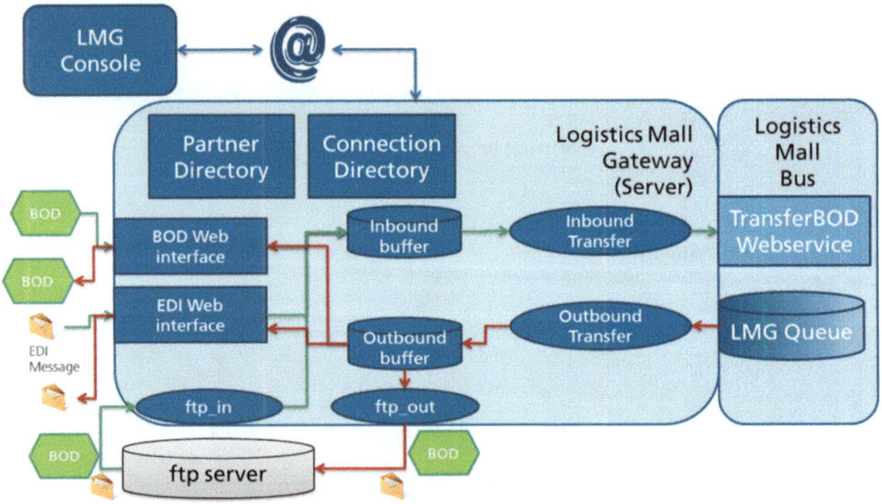

Fig. 16 Logistics Mall gateway architecture

between a product instance in a CAF and registered external parties. Partners and connections are managed in corresponding gateway directories. Each mall customer is enabled to manage communication connections between their booked applications and their external partners. For the information exchange different channels can be used. Currently, web services and ftp is supported, other channels can be added when needed. As exchange format Logistics Mall Business Objects are preferred, other electronic document types can be used as well, but in combination with a converter application in the mall.

The gateway restricts the communication to authorized parties only. For the exchange of Logistics Mall Business Objects the gateway also limits the transfer to object types specified for a defined connection with a registered partner, e.g. partner A may only send purchase orders per web service to the order management system in a CAF if the partner is listed in the gateway directory and the connection is defined there accordingly. For each business object the gateway also checks its structural correctness (Fig. 16).

The Logistics Mall Gateway is implemented by means of the 4D rapid application development environment [10]. The Logistics Mall runs a single gateway server and each CAF gets its own LMG console that is connected with the mall client's part of the gateway server.

6 Conclusion

The Logistics Mall architecture covers the demands of three different stakeholders: end users, suppliers of logistics apps, and mall operator.

End users have a high demand for security besides strong demands for flexibility and availability. That is taken into account by providing an end user with a company exclusive usage environment for apps and processes. Asynchronous communication between apps in the mall and external parties via defined connections and for well-defined business objects increase the security level significantly. Only valid objects from identified sources can pass the gateway, using secured communication channels. The exclusive use of the mall infrastructure assures high availability and performance as side effects from other mall clients are excluded. It also eases the exiting from the mall as client data can be easily returned to the mall customer and can be easily removed in the mall as well.

Suppliers of apps are provided with a defined test environment for checking the mall compliance of their apps. The communication interfaces provided by the Logistics Mall ease the communication of business objects.

The Logistics Mall basic services for accounting, reporting and trouble ticketing also support the supplier business so that he can focus on app development and product management as his core competencies.

The mall operator benefits from the self-administration functionalities of the Logistics Mall Gateway and of the CAF. The client exclusive usage of the mall infrastructure eases the control and management of service level agreements.

While the Logistics Mall is operational in stage 1 for pilot customers, the next development steps still have to prove its viability in practical use. This holds in particular for stage 3 with the composition and deployment of logistics processes.

References

1. The Open Group: TOGAF™ Version 9. Van Haren Publishing, Zaltbommel (2009)
2. Daniluk, D., Holtkamp, B.: Logistics Mall—A Cloud Platform for Logistics. In: Cloud Computing for Logistics, Lecture Notes Logistics, Springer (2015)
3. Weißenberg, N., Stemmer, M.: Modern IT-Platforms for Business Process Management and Portales—Comparative Evaluation of the Tool Suites from IBM, IDS Scheer & SAP, and intalio & liferay. Fraunhofer ISST, Dortmund (2009) (in German)
4. Weiß, N., Leveling, J.: Business Apps Meet the Challenge of Covering Continually Changing Logistics Requirements. In: Cloud Computing for Logistics, Lecture Notes Logistics, Springer (2015)
5. Böhmer, M., Schmidt, M., Weißenberg, N.: Seamless Interoperability in Logistics: Narrowing the Business-IT Gap by Logistics Business Objects. In: Cloud Computing for Logistics, Lecture Notes Logistics, Springer (2015)
6. Bochon, I., Ivens, V., Nagel, R.: Challenges of Cloud Business Process Management. In: Cloud Computing for Logistics, Lecture Notes Logistics, Springer (2015)

7. OAGi: Business Object Document (BOD) Message Architecture for OAGIS Release 9.+. OAGi White Paper Document #20110408V1.0. http://www.oagi.org/oagi/downloads/ResourceDownloads/2011_0408_BOD_Message_Architecture_V9x.pdf (2013). Accessed 12 April 2013
8. Talend ESB: http://www.talend.com/products/esb (2013). Accessed 12 April 2013
9. Java Messaging Service: http://www.oracle.com/technetwork/java/jms/index.html (2013). Accessed 12 April 2013
10. 4D: http://www.4d.com/ (2013). Accessed 12 April 2013

Business Apps Meet the Challenge of Covering Continually Changing Logistics Requirements

Julian Eggemann, Jens Leveling and Norbert Weiß

Abstract This paper describes a business app approach that is used within the Logistics Mall to develop and to execute autonomous IT services. These services are called "Apps" because they focus on a single business function. In this paper the three different kinds of business apps and the architectural components are described based on continually changing logistics requirements. The key aspect of this article is the Workbasket component for apps, including a graphical user interface to manage a task list.

Keywords Logistics Mall · Business apps · SOA · Services · Logistics Mall Apps · App workbasket

1 Introduction

Today a typical logistics process runs in a heterogeneous IT environment. Such systems are mostly standard software like SAP, Oracle, or more specialized logistics software systems like Warehouse Management Systems (WMS) or Production Planning Systems (PPS). All these systems do not entirely meet the requirements of logistics customers regarding short contract duration, pay-per-use accounting and the provisioning of individual IT services. The majority of the features of standard software is rarely used or not used at all, but has to be paid for, due to a monolithic architecture. Additionally, the unneeded functions are often

J. Eggemann · J. Leveling (✉) · N. Weiß
Fraunhofer Institute for Material Flow and Logistics IML,
Joseph-Von-Fraunhofer-Str. 2-4, 44227 Dortmund, Germany
e-mail: Jens.Leveling@iml.fraunhofer.de

J. Eggemann
e-mail: Julian.Eggemann@iml.fraunhofer.de

N. Weiß
e-mail: Norbert.Weiss@iml.fraunhofer.de

M. ten Hompel et al. (eds.), *Cloud Computing for Logistics*,
Lecture Notes in Logistics, DOI 10.1007/978-3-319-13404-8_5

running in the background and consuming resources. In the end, this leads to more unnecessary costs. Finally, customizing software to adjust according to new or changed requirements involves high efforts, costs and risks. Nevertheless, the solution's flexibility for further adoptions is not increased [1].

Apps are the answer to these requirements. The Business App Approach presented in this chapter is based on Service-Orientation as described by [2] and [3] combined with the specific logistics requirements as shown in the next sub-chapter. In order to compose complex processes using existing applications there is a need for modularization, not only as an architectural aspect of a monolithic application but for the whole application. Business Apps are a fine grained approach to reflect the needs of individual processes in the logistics IT. The key feature of an App is to deliver only a single business function.

Finally the composition of these fine grained Apps to a process enables a running business, supported by IT solutions that are covering all individual requirements. Furthermore, the composition of Business Apps allows the adjustment of modelled processes due to continually changing logistics requirements without the ballast of unneeded service functions as pointed out by [3].

2 Logistics Requirements

Logistics often focuses on minimizing the costs over several processes. In this branch process optimization and cost reducing methods like two-stage order-picking are common. This method separates the order-picking-process into two different sub-processes. During the first sub-process all articles for all orders are collected from the stock. In the second step, all articles need to be separated and attached to the given orders.

The Logistics branch requires highly available and easy to use software solutions to resolve complex process chains. The additional requirement is an open interface structure to allow partner companies to exchange data. Furthermore, the logistics companies depend on efficient and stable workflows which need to support by flexible software solutions to resolve the new workflows or process changes.

In general, within a cloud process three different kinds of tasks can be distinguished. The first is a converter service for the integration of external systems. All external systems have their own data structure and are only able to communicate with general data specification within the logistics domain, like EDIFACT [4]. This data has to be converted between the given specification and the Logistics Mall internal business objects (BO) in both ways of communication. The BOs are logistics domain specific objects, like a picking list or a handling unit (HU). The BOs are transport objects, which allow the transport of data in-between the running application or between different apps. For each required data specification a single converter app is required related to the one app one function principle. Integration of external data is needed to migrate existing data into the cloud environment and to connect cloud services with systems hosted on premise. These services are executed autonomously. The second service kind is for executing autonomous business functions like calculations and does not require user interactions. Those functions are common in the logistics domain. Finally the third service kind is for providing a user interface. For each of these kinds a special type of app is defined. The three apps, converter-, service-, and end-user-apps, are described in detail in chapter "Empirical Qualitative Analysis of the Current Cloud Computing Market for Logistics".

For all these kinds a dedicated Business App has to be developed. The "one App one business function" principle as mentioned in chapter "Logistics Software Systems and Functions: An Overview of ERP, WMS, TMS and SCM Systems" addresses the continually changing requirements of the logistics domain. If the customer's requirements have changed, a process could be easily adjusted by removing, replacing or adding different apps. Another benefit of this granular approach is that a customer has to pay only for functionality that is really needed and executed. Furthermore, this rule may leads to small services, which are easier to monitor and to manage as complex services with multiple functions. Additionally, these small services can be exchanged with other services in a running process.

Due to the Service-oriented approach Business Apps are loosely coupled and should have no knowledge about other services or environment specifics. They have to provide an interface that is able to accept Business Object Documents (BOD) and to send BODs to the Process Engine's Message Broker. A BOD can be described as an envelope that contains a list of BO instances of one type. The BOD itself contains information about the origin of the BOD. BODs that are exchanged with the Process Engine are only containing BO instance IDs. The BO instance details have to be read from the BO Instance Repository by the App themselves. The Broker is defined as central bus infrastructure for communicating between Logistics Mall environment[1] essentials. Furthermore, Apps have to request the required BO instances, which are presented by their IDs within the BOD, from the BO Instance Repository. These three ways of communication must be provided and implemented by each Business App that should be executed within the Logistics Mall Cloud environment.

The Business App architecture should also be technology neutral to enable the adaption of already existing IT services for the Logistics Mall. Those services are based on several technologies. The lowest standardized common denominator of in the Logistics Mall is used by the environment. The communication infrastructure should be designed with widely accepted standards as defined by [2]. This enables a wide range of technologies a Business App could be developed with.

3 Logistics Mall Specifics Requirements

In this chapter requirements of other Logistics Mall components are described. The focus is here mainly on the process engine that interacts as mentioned before with the apps.

[1] More details in chapter "The Logistics Mal—An IT-Architecture for Logistics-as-a-Product".

The process engine is not able to differentiate between App types. Additionally, the Business Object Instance Repository is not able to lock BO instances due to its client wide usage. Further, the repository does not have any knowledge about Apps or other clients, like the process engine. It is just a system wide data storage approach. To consider this issue, End-user Apps have to provide a mechanism to store incoming BOs until a user completely processed it. This mechanism has to avoid errors due to parallel processing of one BO. A BO has to be locked directly, if a user started to process it.

4 Business App Architecture Approach

This chapter brings the Logistics requirements and the Cloud environment together and describes how our Logistics Mall Apps cover these requirements by an application view. For a better overview, this chapter names requirements and presents the cloud environment. Later the Logistics Mall App structure and the different App architecture layers are described. The mentioned requirements can be realized by cloud solutions, such as the Logistics Mall. The following paragraph describes the cloud environment used by the Logistics Mall Apps.

The Logistics Mall offers a cloud environment to fulfill the above mentioned logistics requirements. The Logistics Mall represents a marketplace for logistics software and services which can be accessed over the web-front-end designed by the cloud operator. The designed architecture based on the Software as a Service (SaaS) service model. In this model software systems are executed in a cloud environment and accessed through a Web Browser as described by [5]. Furthermore, the cloud infrastructure is managed by the cloud operator. Each customer can book the required software apps. Before use, the booking needs to be approved and provisioned. Afterwards, the customer is able to simply use the booked applications over the web-browser. The advantage of the mentioned purchase process is based on the flexible use of the booked software solutions running on cloud solutions, such as the Logistics Mall. Based on the flexible use, the apps are offered on a pay-per-use billing. Next to the pay-per-use billing option, other product subscriptions can be offered, such as a "full service"—package. The way of billing is arranged by the application provider. A main advantage of the Logistics Mall is based on the creation of individual logistics processes. This is done by the Logistics Process Designer (LPD).[2]

Moreover, the service providers can offer their services to customers by adding their Apps to the Logistics Mall Marketplace. Before the applications can be used within cloud environment, the service providers need to adapt their software applications to the standards, employed by the Logistics Mall. In addition to that, the App has to be adapted to the cloud computing solution.

[2] More details are available at chapter "Challenges of Cloud Business Process Management".

After introducing the Logistics Mall and the basic idea of Apps being used in the Logistics Mall, the following paragraph focuses on the structure inside the cloud infrastructure. The major difference between local software solutions and cloud-based software solution is the different delivery of the functions. While local software needs to be installed on local hardware, the cloud-based solution is provided as a service on the external environment. An environment for Business Process as a Service (BPaaS) solution requires a more complex infrastructure to support a high number of customers.

A booked app is provided as an instance to every customer. Each instance could be used within multiple processes. Multiple instances of one app will only be created, if an instance is highly used and more memory and processing power are required as the app or the app execution environment can manage itself. Additionally, the Logistics Mall delivers for each version of an app a single instance, if an App is needed in different versions. The company's data, as well as the booked instances, are stored in the cloud-environment. Due to privacy and security reasons, the data is separated from other customers' data by offering an individual platform per Logistics Mall customer. The data is stored in a general database, called the BO-Instance-Repository.[3]

4.1 App Instances

An App could be deployed on several Application servers. Each deployment of an App within the Logistics Mall is named App Instance.

4.2 Layers

The following paragraphs describe the Logistics Mall App structure including the different layers of the used structure. An App running on the Logistics Mall cloud platform is divided into three layers as shown in Fig. 1. The layers are based on the Model-View-Controller (MVC)—Pattern as described by [6] extend by the Workbasket that is described later on. The Model of each App are the used BOs.

4.2.1 Graphical User Interface (GUI)

The first layer realizes the interaction with the clients running the application. This is implemented by a web-UI to allow flexible use on most devices such as mobile

[3] More details are available at chapter "Seamless Interoperability in Logistics Narrowing the Business-IT Gap by Logistics Business Objects".

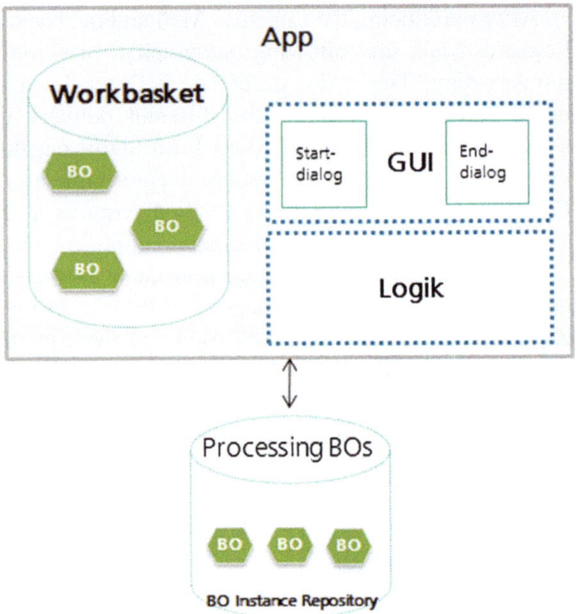

Fig. 1 Logistics Mall App Layers

phones, tablets and computers. The user will be guided through most processes. The dialogue usually begins with the start dialogue (start section) and finalizes the app-process with the end dialogue.

4.2.2 Business Logic

The business logic interacts with the GUI and represents the algorithms and logic behind each application. All functionality is run and organized by the business logic while the transfer of data is structured by standardized objects, known as the business objects (BOs).

Additionally, the business logic can retrieve business object data (processing BOs) from the BO Instance Repository to achieve functional goals which cannot be reached with the information given from the incoming BOs.

4.2.3 Workbasket

The Workbasket approach, as explained in this sub-chapter, addresses the requirements for a mechanism to store all incoming messages from the process engine and for locking these, if a user started the processing of a BO. The main duty is to store all incoming Mini-BODs until a BO referenced by one BOD is completely processed. As shown in Fig. 2 there is one Workbasket instance for each executed app instance that is shared through all user sessions of this app instance.

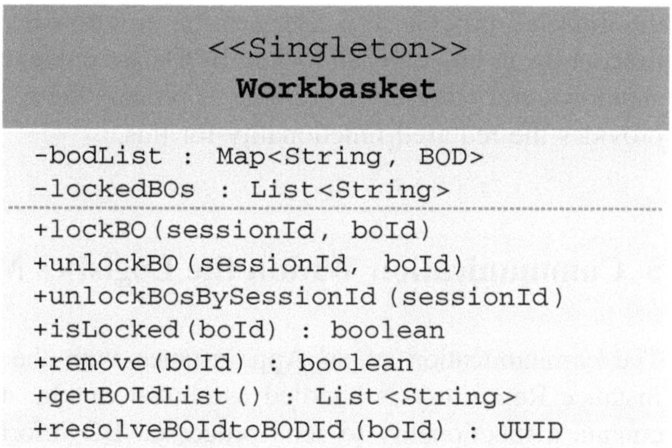

Fig. 2 Workbasket singleton class diagram

The Workbasket contains also methods for locking and unlocking the BOs within the received BODs. Every web user interface instance has its own unique id, which is used to lock Mini-BODs, this disable the parallel processing of one BO instance by two or more users.

The Workbasket is the part of the presented approach, which enables the split and join of process instances. For example, an order (the input BO instance) with four positions is split into two pickings lists (output BO) by a Business App. The output of the App are two single BODs containing each one of the picking list BOs. The next App instance within the process processes both BODs separately, but is able to add parts of both pickings lists to one new delivery. All the splits are leading to new process instance to handle the separate BODs.

A new entry to the Workbasket BOD-List is automatically added if the APP instance RESTful interface is called by the Process Engine. The App approach is designed for Java Enterprise Edition Version 6. The adding of a new BOD is handled by the Context and Dependency Injection (CDI) event mechanism. That allows a loose coupling between the RESTful interface and the business logic and GUI of an App Instance Fig. 3.

After a BO instance is completely processed, it should be removed from the Workbasket. Additionally the Process Engine will be informed by message about the by instance process finalizing. Afterwards, the process engine is able to continue the process instance by sending the results of the app instance to the next app along

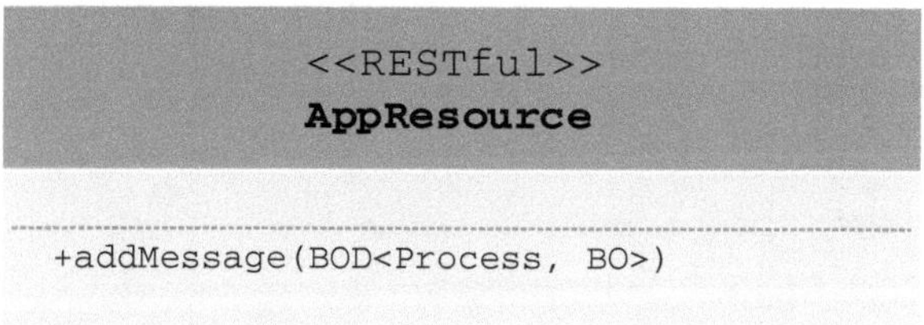

Fig. 3 Workbasket interface

the modeled process. If a user session is terminated due to an error or a session timeout by an idle user, all locked BODs are unlocked automatically by the session shutdown and other users are able to process the released BODs. The Workbasket provides the required functionality for this.

5 Communication Within the Logistics Mall Infrastructure

The communication of an App instance with the Process engine and the BO-Instance Repository is handled as shown in Fig. 4. The different Apps have to contain a functionality to send processed BO information to the Process Engine. Business Apps have to provide an interface, which is called by the process engine to send new jobs to an App instance. The functionality and message characteristics are described within this sub-chapter.

The communication within the cloud environment is implemented by a RESTful API. A main advantage in using REST inside the Logistics Mall is the HTTP-based structure, which allows an easy transfer of information throughout the whole cloud-environment. Furthermore, REST is a lightweight and easy to use implementation compared to other projects [7]. The Logistics Mall uses Business Objects (BOs) to transfer information over REST, through the supported types XML and JSON.

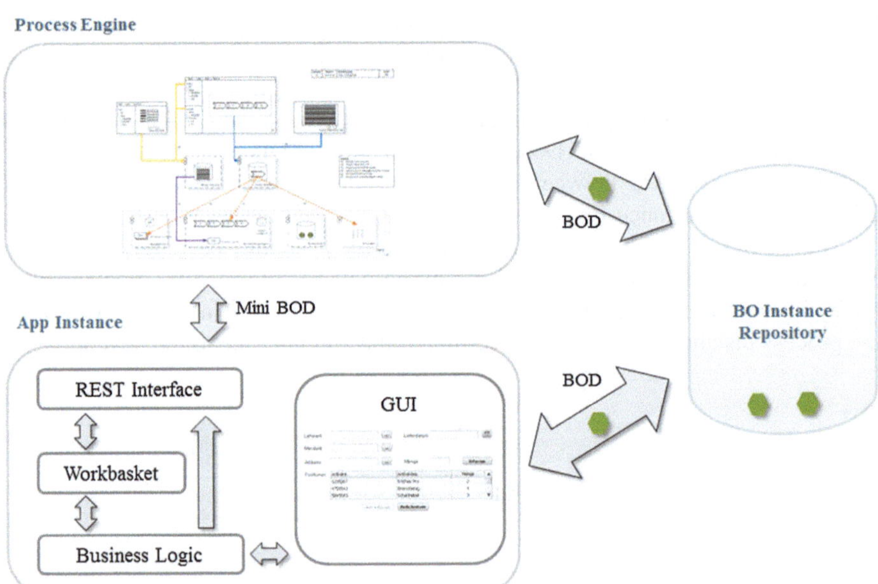

Fig. 4 Architecture from App Instance perspective

More precisely, an app is transferring information based on a defined structure. This structure is specified by BO-Definitions. A BO instance is transferred within a Business Object Document (BOD). A BO Instance contains all necessary information about a real domain specific object like a shipment or a handling unit. The Business Apps receive minimized BODs from the Process Engine. These messages, called "Mini-BODs", contain only an ID of a BO instance. This is part of the interface provided by the App. That means the App describes the BOD-type that could be processed. All other types are not accepted.

The App has to load this instance. Afterwards, an App is able to process the loaded BO for further operations within the given workflow.

5.1 App Characteristics

This chapter outlines the common structure of apps deployed in the Logistics Mall infrastructure. The applications can be categorized in three different groups. Each of the groups is presented below.

5.1.1 Converter Apps

The applications categorized in the "converter app"-group deploy the functionality to translate the application-data of external Systems into the BO standard. In fact, the application-data is converted to the equal Business Object used in the Logistics Mall to deliver the given information in a valid format. Furthermore, the Converter App also allows converting the BOs used in the Logistics Mall to the valid format of the connected software solution. Every Converter App accepts exactly one format. Communication with external systems is only possible through the Logistics Mall Gateway.[4] Additionally, converter Apps receive messages not solely from this gateway, but from the Message Broker. All incoming external Messages are sent to the Message Broker first due to security considerations Fig. 5.

5.1.2 Service Apps

The second category of apps is known as Service Apps. The main purpose for Service Apps is to deliver one business function to the Logistics Mall environment. For example this could be a load space optimization algorithm. These apps do not provide a user interface or contain a workbasket. The service delivered by a Service

[4] See chapter "The Logistics Mal—An IT-Architecture for Logistics-as-a-Product" for more details.

Fig. 5 Converter Apps

App will be automatically executed, if the provided RESTful interface is called by the process engine.

The RESTful interface of a Service App has the same signature as the already described workbasket interface, but no BO Ids are be administrated Fig. 6, 7.

5.1.3 User Apps

The last group of applications is summarized as User Apps. A User App describes the interaction with different clients. While traditional software solutions can be used over Graphical User Interfaces (GUI) to accomplish tasks, the Logistics Mall User Apps offer a simple access to the user interface via a web-front-end. Over the Web-front-end the user can simply run the functions due to his level of restrictions (his role). Finally, the User Apps representing the only interaction between a user and the cloud-computing service.

Fig. 6 Service Apps

Fig. 7 User App

6 App Life Cycle

The Business App Life Cycle is divided into five steps. The following Fig. 8 describes the workflow within the App Life Cycle. The five steps are detailed within this chapter.

6.1 Development

The first step is the development phase. An App Provider implements a service following the guidelines described in chapter "Empirical Qualitative Analysis of the Current Cloud Computing Market for Logistics". The hard requirements are that RESTful interfaces and service calls must be implemented. Additionally, the BO-stack including BODs and Mini-BODS, of the Logistics Mall environment must also be used for communication and the BO Instance Repository must be used for storage of processed information and data shared by different apps of a process. Furthermore, an end-user and the service App has to contain the workbasket mechanism. Additionally, points are just suggestions to the provider, like the usage of the Java enterprise stack. The developers are free to choose their own programming language, but must make sure that their apps are executable within the cloud environment. This is ensured and verified during the next phase of the Logistics Mall App Life-Cycle. The development phase finishes with submitting the created App and integrating it into the Logistics Mall Marketplace (MMP). For the integration the app's description, its price model and the date of availability are registered in the MMP. A Business App is only available until the specified date. But first of all the App is not visible or purchasable for any customer as long as the Logistics Mall Verification has not been successfully completed.

Fig. 8 App Life-Cycle

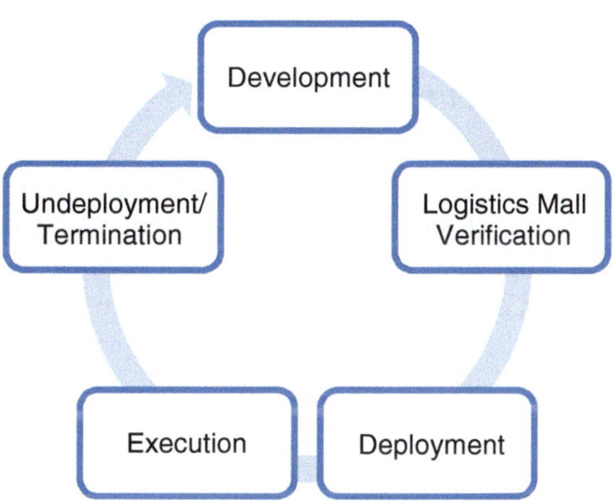

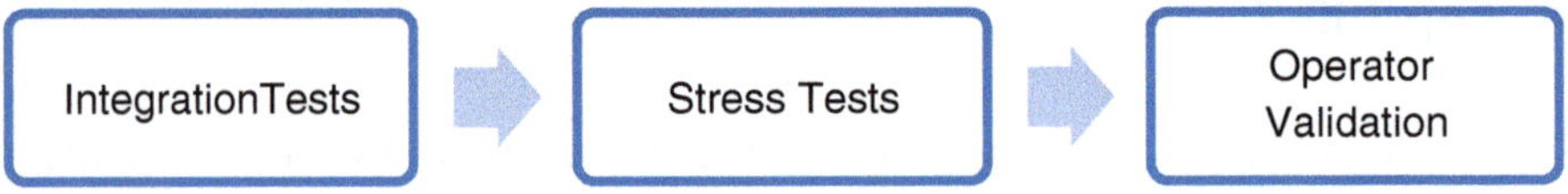

Fig. 9 Verification phase

6.2 Logistics Mall Verification

All Business Apps have to be verified. Apps are tested to make sure that they are covering the hard requirements. Also the tests verify a stable execution of Apps. Additionally, apps are analyzed for malware[5] detection. The main focus of this phase is the stable execution within different processes. This is tested and validated by stress tests and integration tests Fig. 9.

During the integration tests an app is deployed and executed within a test environment. For the stress test an App instance is called sporadically with both correct and erroneous messages. The erroneous messages are not compatible with the RESTful interface of the App, they do not have the right format or they are not even complete. This kind of stress test ensures a stable execution of any Business App in the Logistics Mall environment. This phase is executed by the Logistics Mall operator. If the test fails, the app provider has to do correct the occurred errors and requirement lacks. If successfully verified the App is visible within the MMP and entitled to be purchased and used for process modeling by any customer.

6.3 Deployment

Once a process is purchased by a customer, the operator starts the App's deployment in case the app has not been instantiated before. Due to the successful verification this should be done without any errors.

6.4 Execution

The customer executes the new App instance within his processes. As mentioned in chapter. "Empirical Qualitative Analysis of the Current Cloud Computing Market for Logistics" end-user apps are executed by any user of the customer company. Both Converter- and Service-Apps are executed automatically.

---

[5] Malware are programs that are containing malicious operations.

6.5 Undeployment/Terminating

If an App Instance is not used within any process instance, the operator will stop this instance gracefully. Afterwards, the App instance has to be marked as "terminated" in the Logistics Mall Database. Once all instances of an App are terminated and the App availability has exceeded, the App Status has to be set to "expired". The App provider is still able to add a new version of the App. Additionally, the availability could be renewed.

7 Conclusion

The workbasket approach successfully bridges the gap between the needs of logistics applications and the process-oriented view of the Process Engine. This gives the opportunity to develop applications that can overcome the single process instance focus of BPM with the side effect that processes will become more complex due to changes of granularity. In our example of two-stage order-picking this means a change from dealing with picking orders which are each owned by a single process instance to the challenge to deal with picking order positions of multiple picking orders in the first stage. The second stage has to ensure the fulfillment of the initial order by joining the results of the first stage and dividing the intermediate picking positions according to the needs of the original orders. This change of granularity will lead to additional process instances which have to be synchronized. Balancing the aspect of increasing the complexity of the process model versus encapsulating complex functionality in single apps will be a future task.

The presented architecture of the apps has proved to be satisfactory. Easy to understand and maintain for the developer, flexible and lightweight but providing a good framework for implementing many different apps. Granted that there are enough apps to cover all needed processes the goal of providing the desired functionality at lower costs and higher flexibility can be reached in the future.

References

1. Wolf, M.B., Rahn, J., ten Hompel, M.: Cloud computing für logistik 2‘, fraunhofer verlag, ISBN 978-3-8396-0558-5 (2011)
2. Papazoglou, M.: Web Services & SOA—principles and technology, 2nd edn, isbn 978-0-273-73216-7, pp. 778–779 (2012)
3. Erl, T.: Service-oriented architectures, 2nd Edn, Prentice Hall, isbn 978-0-132-344-821, (2008)
4. EDIFACT Standards Overview Tutorial: Learn About Key e-Commerce Trends and Technologies at Your Own Place, GXS, Online, http://www.gxs.co.uk/wp-content/uploads/tutorial_edifact.pdf01 Aug 2014

5. Holtkamp, B., Steinbuß, S., Gsell, H., Löffeler, T., Springer, U.: Towards a logistics cloud, presented at the 2010 sixth international conference on semantics, knowledge and grids, pp. 305 −308 (2010)
6. Elias, S.J., Shahuddin, A.Z., Shaharudin, M.R.: Development of ISE club information system (ISECIS): based on model-view-controller (MVC) pattern, LAP lambert academic publishing, ISBN: 978-3-8443-2211-8 (2011)
7. Reitbauer, A., Novakovic, M.: RESTlos glücklich? in javamagazin, pp. 56−58 (2009)

Seamless Interoperability in Logistics: Narrowing the Business-IT Gap by Logistics Business Objects

Martin Böhmer, Michael Schmidt and Norbert Weißenberg

Abstract Logistics IT is shifting from large, monolithic software systems to smaller, modular applications ("apps") that can be flexibly composed to support individual business processes. In order to anticipate extensive integration projects, seamless interoperability is a crucial requirement. This paper outlines a standards-based logistics Business Object model that narrows the business-IT gap and simplifies logistics IT services regarding development and integration.

Keywords Logistics Mall · Business Objects model · OAGIS · Business-IT gap · Process modelling

1 Introduction

Logistics processes and supply chains are highly dynamic. They usually involve a huge number of parties and rely on a variety of IT systems of different providers.

Establishing and maintaining these processes according to always changing business requirements poses a big challenge on both the business level and the technical level. The traditional separation of (and the gap between) business and IT is an additional obstacle to overcome.

M. Böhmer (✉) · M. Schmidt
Fraunhofer Institute for Material Flow and Logistics IML,
Joseph-Von-Fraunhofer-Str. 2-4, 44227 Dortmund, Germany
e-mail: Martin.Boehmer@iml.fraunhofer.de

M. Schmidt
e-mail: Michael.Schmidt@iml.fraunhofer.de

N. Weißenberg
Fraunhofer Institute for Software and Systems Engineering ISST,
Emil-Figge-Str. 91, 44227 Dortmund, Germany
e-mail: Norbert.Weissenberg@isst.fraunhofer.de

M. ten Hompel et al. (eds.), *Cloud Computing for Logistics*,
Lecture Notes in Logistics, DOI 10.1007/978-3-319-13404-8_6

Flexibility and modularity of (physical) logistics services and IT systems have become crucial aspects for rapid process adaption. Hence, extensive IT systems are increasingly subdivided into smaller software applications (apps). When composed to business processes, seamless interoperability of these apps is a requirement for significant competitive advantage. Considering the business-IT gap, standardisation is necessary on both levels in order to narrow the gap and accelerate process adaptations. This is what Business Objects are about.

The Fraunhofer Institute for Material Flow and Logistics IML and the Fraunhofer Institute for Software and Systems Engineering ISST have developed the communication standard Business Objects for Logistics (BO4L [1]) in cooperation with the Leading-Edge Cluster *LogistikRuhr* [2]. The main objective is to define the first independent standard for the logistics domain, namely BO4L, which enables logistical IT applications from different vendors to seamlessly interoperate (narrowing the business-IT gap).

A Business Object (BO) represents an object relevant to a business domain, such as a product, an order, or a handling unit. There may be a physical counterpart for a BO, as for a product, but also non-physical objects like a pick list can be described. A BO has a structured set of attributes that allows a comprehensive description of the object for various applications.

Business Objects are prerequisite for the implementation of extensive, modular logistics IT solutions—particularly with regard to cloud concepts such as the Logistics Mall [3, 4]. The BO4L standard is already in production use in a German drugstore ordering system [5], presented at the German *IT Gipfel* 2012 [6]. In this context, the Business Objects enable a continuous, standardised communication between an online store, an enterprise resource planning system (ERP) and a warehouse management system (WMS).

BO4L constitute a business-driven domain model and a technical standard for interoperability at the same time. Thus, they do not solely target IT systems or applications on a technical level, but also the communication of the involved parties on a business or analyst level. Merging business and technical aspects, the BOs enable domain experts and analysts to model, deploy and maintain business process chains without significant intervention by IT personnel. This closes the business-IT gaps significantly, and also closes some horizontal gaps, as detailed in this paper.

BO4L is not only a data model with different views, it is accompanied with supporting software: a BO Instance Repository and a BO Type Repository, both with Web browser support.

"Information is the new oil" [7] for the industry and this oil has to be standardized with respect to its types (the BO model) and its de-duplicated instances (the BO Instance Repository, being a central data service). Hereby, development and maintenance costs as well as time-to-market are minimized.

The paper is outlined as follows. Section 2 describes the main characteristics of the Logistics Business Object Model BO4L, including its functional domain model and several technical model representations, as well as procedure models for model design and model governance. The issue of narrowing the business-IT gap by the use of Business Objects and the closing of horizontal gaps is discussed in detail in

Sect. 3. Since BO4L is more than merely a data model with different views, but also supports persistence of instances and types, these supporting software components for the BO model are explained in Sect. 4. Finally, background and related work are presented. The paper concludes with a summary of all achievements provided by BO4L.

2 The Logistics Business Object Model BO4L

We present the objectives for developing the Business Object model, the resulting design and structure of its functional and technical layers, followed by corresponding development methods and the model's governance process.

2.1 *Objectives*

The objectives of a standard Business Objects model are on the one hand to conciliate logistics and IT and on the other hand to mediate between different IT systems (ERP, WMS, etc., see [8]) and parties. The former yields a common understanding of Logistics and IT departments. The latter has two aspects: to support fixed communication of IT systems and to support a flexible orchestration of IT systems using business process models by defined standard data objects for logistics process models, as described in Sect. 3.4.

Hence, the business-IT gap must be narrowed vertically (Logistics-IT) and horizontally (along the process chain). The vertical gap separates the business process and data from their IT representations, while the horizontal gap comprises an IT-to-IT gap and a Business-to-Business gap. This is illustrated by Fig. 1.

These objectives result in various requirements for the domain model and its technical representations. A key requirement is the use of standards as described below. Finally, the requirements are summarized by our definition of the term Business Object.

2.1.1 Requirements for a Business Object Model

Targeting the vertical business-IT gap, the BO model would ideally merge the two sides, so that there is single representation only. But in this case merging also implies mixing of aspects specific to each of the sides and not helping the other side. Instead, the model should consist of an abstract, domain-driven model (domain view) with one or more technical implementations, each extending the abstract model by its specific characteristics. The domain view and its technical representations (referred to as technical view) must be coupled tightly to effectively bridge the business-IT gap. Both views have to be extensible, versioned synchronously and controlled by a

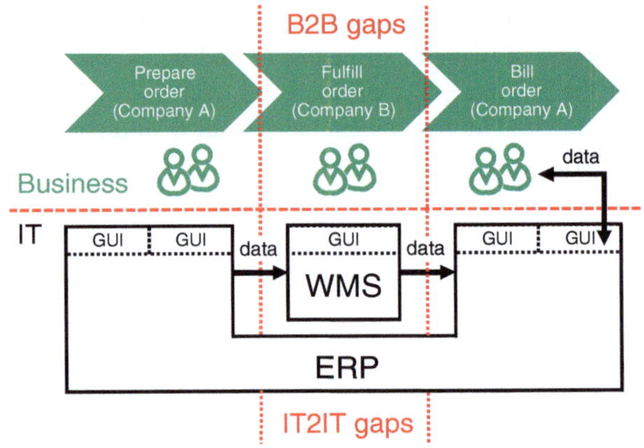

Fig. 1 Vertical and horizontal business-IT gaps

governance process, due to the complexity and dynamics of the (logistics) domain. Different model versions must not interfere with each other in a process execution environment when used simultaneously, e.g. by different processes.

The **domain view** abstracts from any technical details and can easily be understood by domain experts and business users. It specifies relevant physical objects (e.g., products, palettes, inventory) and virtual or logical objects (e.g., pick lists, order lines, storage areas) and their relations by reflecting the real world (domain).

The set of BOs must form a domain model enabling a consistent description of comprehensive logistics processes. This includes defining (business) entities by means of reusable, structured components, basic attributes (e.g. numbers, codes, strings) and relationships to other entities and components. For ease of understanding these definitions should be provided in a graphical representation based on a standard notation like UML. The domain model should be canonical (i.e., having simplest archetypical structure) and based on or compatible to existing standards as far as possible. Respecting established standards and extending them is the key for broad acceptance. Communication using BOs may be legally binding, requiring the specified entities to represent business transactions.

The **technical representations** should also be standards-based and enable seamless integration on different IT layers for realising short time-to-market. The ultimate goal is to support simple and flexible composition of apps from different vendors. Thus, the technical representations of BOs are employed to specify the app's interface profiting from the semantically precise and vendor-independent domain model. BOs are the key to app exchangeability, meaning an app can be replaced by another app as long as the in- and outgoing BOs and the app's semantics match. In combination with the domain model, the technical representations/views of BOs must be suited for creating exchangeable application building blocks (apps). Finally, the technical views have to simplify app development in a way that enables app developers to focus on their business logic.

The BO model is the key to avoid unwished dependencies. It does not only decouple domain layers and technical layers—at the same time it connects them by being a common language for both: The BO model decouples higher domain-oriented architecture layers (especially business process models and their modeling tool and execution engine) from lower technical layers (especially services and applications). This increases the flexibility and agility of adapting process models and of exchanging service implementations, since the BOs are *not* service-specific, but provided similarly by any substitute service. This is explained in detail in [9], pp 82–86. For example, a one-step order-picking app can be replaced by a two-step picking app without affecting the rest of the process.

The structure of BOs is the same for different use cases. However, different use cases have different constraints: When using BOs for *query by example* (resulting all objects that are similar to this sample) or for simple unmarshalling (e.g. converting XML to Java), BO attributes must be allowed to be empty. However, when communicating them to partners or when storing them, they have to fulfill all constraints defined by the domain. This also applies for using BOs with different verbs or methods. The constraints a BO has to fulfill also depend on its state: a BO has a lifecycle, and in each state, other attributes of the BO are required, others are optional. Therefore, the validity of BOs is a situation aspect and not an integral part of a BO definition.

2.1.2 Benefits of Using Standards

BO4L was intentionally designed to harmonize with existing standards. Open, accepted standards result from the condensed experience and best practice of the standard development team.

Using such standards contributes to the standardization of interfaces and products, as e.g. detailed in [10] and [11]: Unified information in the form of Business Objects used across the entire value chain enables reuse and exchangeability of (physical and/or business) services and software modules, ultimately cutting costs, reducing risks and complexity as well as increasing data quality.

Providers (of services or software) can target larger markets, while users face lesser risk of vendor lock-ins and profit from intensified competition among providers. Both providers and users gain investment security.

Moreover, lower transaction costs result, since standards accelerate business processes, shorten negotiations, and simplify contracts. Standards for technical interface parameters between subsystems enable an efficient construction of complex systems and lower development costs, especially in cases involving a lot of sub-systems, frequent changes and agile development.

2.1.3 Business Object—A Definition

Business Objects specify real or virtual entities relevant to a particular domain by structured sets of attributes, reusable sub-components and relationships to other BOs. They serve as standardised means of communication for business processes and the IT systems these processes rely on. Thus, BOs are suited for specifying interfaces on a business *and* on a technical level to facilitate Business-to-Business, IT-to-IT and Business-to-IT integration. A BO has an abstract domain view and one or more technical representations [12].

A BO model (in domain view) reflects a domain model enabling a consistent description of comprehensive business processes in that domain. It abstracts from all technical details not relevant to business users or domain experts. E.g., in a business process model usually only the names and states of BOs and a few control flow-relevant attributes will appear, as explained below.

A BO in technical view, i.e. a technical representation of a BO, is a software artefact that implements the corresponding domain object with all relevant technical details required for a particular purpose or technology. Thus, a BO exists in various inter-convertible technical representations, e.g. for communication (transport) between systems and system components (applications, services, infrastructure components, process models) or for persistence. A BO's technical representation is also employed for the passive control of processes and involved systems: decisions are based on the BO's state and attribute values and the BO basically does not actively influence its environment.

It is important to distinguish between BO types and BO instances. While a BO type serves as a blueprint used when designing processes or software (e.g. "customer"), a BO instance denotes an actual, "living" object build according to a BO blueprint (e.g. "Acme Ltd.").

2.2 Design of the Business Objects Model

To close the horizontal gaps only requires a common canonical model on all layers. To close the business-IT gap, the BO model is designed to be used by logistics domain experts as well as by IT-experts and system integrators.

However, since these groups have different focus, they are supported by different views of the BO model. To narrow the business-IT gap, these views are tightly coupled and automatic translations are supported, as depicted by Fig. 2.

2.2.1 Domain Model—Functional View

As described above, the domain model is used by logistics experts. In order to be easy readable and accessible and also transferable into different representations (e.g. Java), it is represented as UML class diagrams.

The Domain Model describes different objects (including their attributes) that occur within the logistics business respectively that are communicated within logistic processes. This includes information about physical objects (e.g. articles or load support) and documents (e.g. orders).

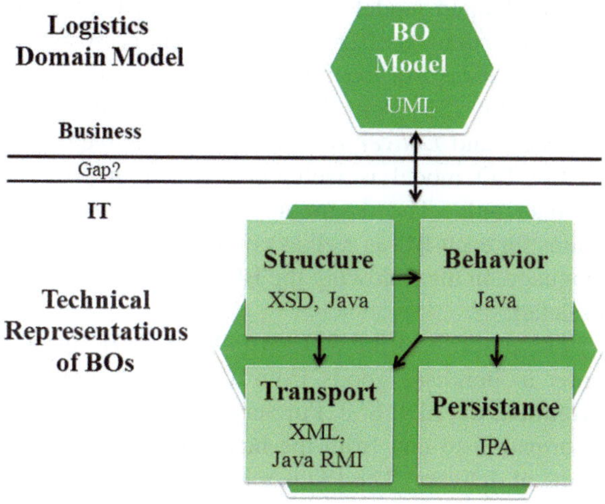

Fig. 2 Business object model and Views

BOs are composed of attributes and components, where many components are standardized, e.g. by CCTS [13] and other UN/CEFACT [14] standards. This holds for OAGIS [15] (Open Application Group Integration Specification) as well as our Business Object model. Components themselves are also composed of attributes and components, and from so-called reference components other BOs can be linked. Therefore, components are the standardized building blocks of BOs. This is displayed by Fig. 3.

This also holds for the Java representation and the database representation of BOs, but here references between BOs are represented naturally via BO-typed attributes or foreign keys respectively.

The bandwidth of the BO attributes is ranging from features like a unique number to notes and descriptions, physicals dimensions, various statuses etc.

Currently, the domain model consists of 31 business object types, covering the following domains:

- Order Management: from purchase order to invoice
- Facility Logistics: from receiving to shipping
- Supply Chain and Transportation: inventory, capacity and route information.

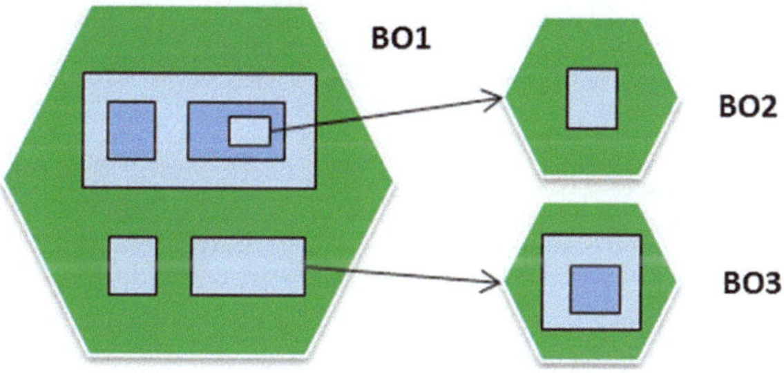

Fig. 3 Inner structure of business objects

The domain model consists of more BOs than found in the UML model: The domain model adds BOs, which only differ in state or verb/method usage, but have the same form. An example is BO *Shipment*, which is also used for *NotificationOfDispatch* (*Shipment* [announced]), *Fulfillment Confirmation* (*Shipment* [dispatched]) and *DeliveryNote* (*Shipment* [delivered]).

The BO model is well documented in [16], however currently available in German only. The documentation provides motivations and detailed UML diagrams for each BO as well as comments for each attribute in German and in English language. In the following the BOs *PurchaseOrder* and *HandlingUnit* are explained in detail.

The purpose of the *PurchaseOrder* (PO) business object is to communicate an order to purchase goods from a buyer to a supplier (see Fig. 4). The PO is an evolution of EDI 850 Purchase Order and OAGIS *PurchaseOrder*. It carries information to and from the buyer and supplier, including terms of payment and terms of delivery. The PO is a legally binding document once both parties agree to the contents and the specified terms and conditions of the order.

A *PurchaseOrderLine and a SalesOrderLine* are linked with a *WarehouseShippingOrderLine*. The *SalesOrder* (SO) is a step beyond a PO in that the receiving

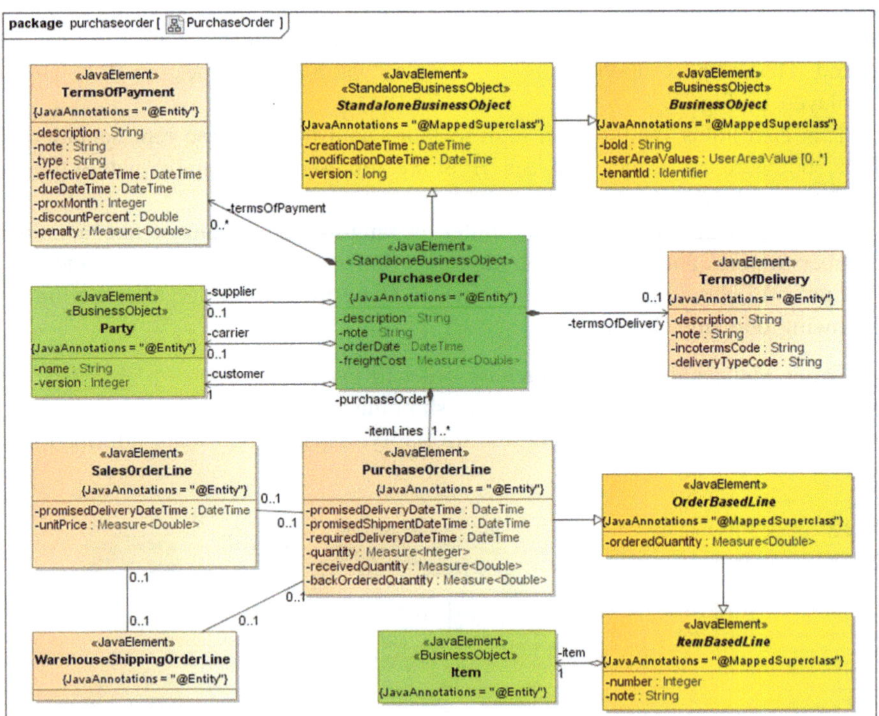

Fig. 4 Business object purchase order

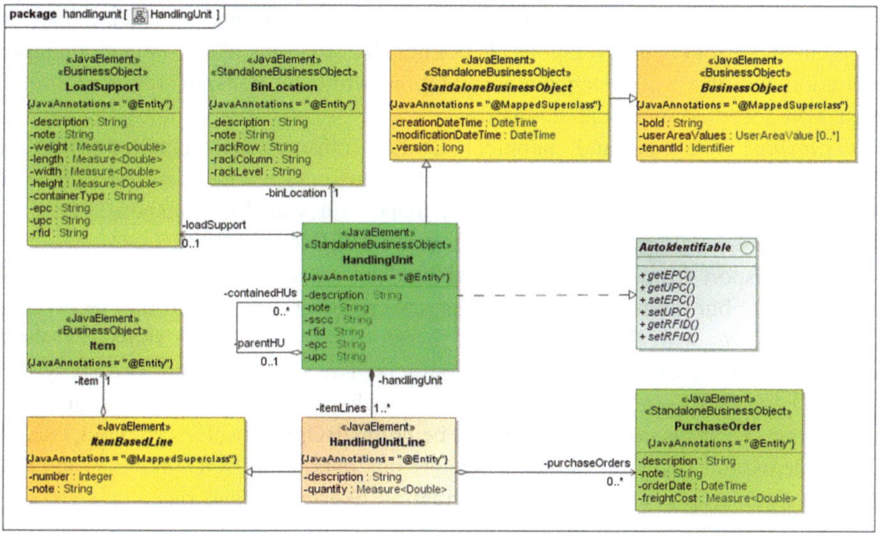

Fig. 5 Business object handling unit

entity of the order also communicates sales information. The *WarehouseShipping-Order* (WSO) is used to instruct warehouses to ship orders. It is commonly used to authorize a warehouse to make, confirm, modify, or cancel a shipment or a previous shipping order. By associating order lines, partial deliveries and combining of orders in the transition from one order to another one can be realized.

The understanding of the BO *HandlingUnit* (HU) follows the definitions of [17] and [18] (see Fig. 5). A HU is a physical unit that consists of load support and goods contained therein or thereon. HUs are the basic units of all material handling processes (put to stock, retrieve from stock, order picking etc.) and associated informational processes (scan/identify, inventory control, etc.). All the information contained in the items, such as batches and serial numbers, location where the items are stored, physical dimensions etc. are available by a HU. HUs can be nested, meaning that one HU can be created from other HUs (e.g. HU Container → HU Pallet → HU Case). They have unique identification numbers that can be formed according to standards such as GS1-128 or SSCC (Serial Shipping Container Code) [19].

2.2.2 Technical Model Representations

Each BO technically comes in different representations, which technically enable or focus on different aspects and which are designed to support different architecture layers. Each of them is machine-interpretable and derived from the logistics domain model (UML), since we generate Java code from the UML model, and Java is central for the generation of all technical representations.

Structure. The structure is defined by the UML domain model, which names the BOs, their attributes and relations. It defines the structure of the behaviour representation and of the persistence representation, while the transport representation is based on accepted technical standards.

Transport. The main and canonical standard transport representation is OAGIS [15], namely our light OAGIS overlay and its business object documents (BODs). However, by using converters, any other market-relevant BO transport format can be used if a legacy application needs it, e.g. UN/EDIFACT [20].

The transport format is transactional: It does not only consist of data (one or more BOs), but additionally describes the receiver and what the receiver is requested to *do* with the data. OAGIS therefore provides BODs (business object documents) being exchanged between partners. The BOD structure is displayed in Fig. 6: Each BOD mainly consists of a verb (what to do with the BO) and one or more nouns (or BOs, as we call them). The nouns/BOs themselves are composed of components and fields (attributes). Besides this data area, BODs have an application area containing some general information like sender and receiver.

The verbs express what the receiver is expected to do with the nouns. The verbs come in pairs (request/response) and are also standardized by OAGi. They can in principle be combined with all nouns/BOs, which reduces the learning curve for BODs. *Get* and *Change* represent CRUD methods; *Process* is used for business logic. *Notify* means using BOs for events and asynchronous communication. *Sync* is used for master data synchronization. Figure 7 displays this orthogonal construction.

BOs in transfer format separate state and structure from their behaviour, because they are communicated between different systems (and possibly also between different layers and components of one system). This holds because behaviour is not transferable nor easily portable and has to be coordinated flexibly by higher-level business process models. Emergence (interference of different behaviour descriptions) should be avoided.

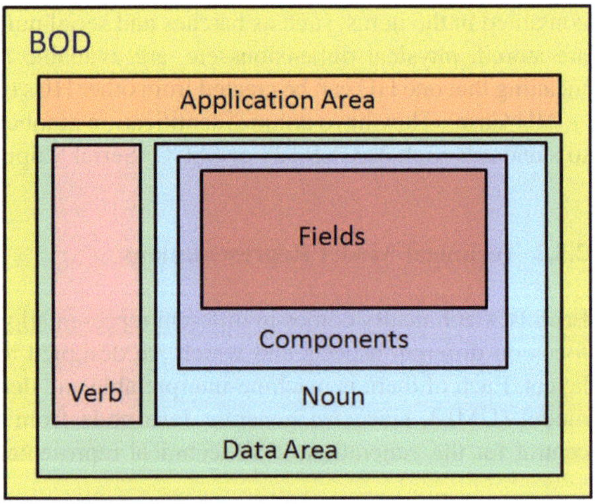

Fig. 6 BOD structure (from [21])

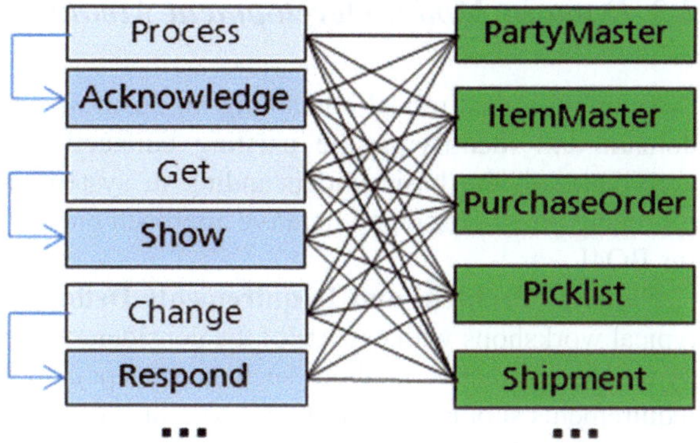

Fig. 7 Verbs are paired and freely combinable with nouns/BOs

Behaviour. Business Logic is supported by the Java representation of the BO model. A Java entity model is generated from the UML model. It provides CRUD methods and more. Thus behaviour information is type information, not transported with individual object instances.

Persistence. The persistent representation of the BO model is generated from the Java representation using JPA (Java Persistence API). In this way, different relational database system and even NoSQL databases can be used to store and manipulate BO instances. We even provide a BO Instance Repository (see Sect. 4.1).

A BO therefore exists in different inter-convertible technical representation formats: the standard transfer format may be parsed into equivalent technical objects (with methods in their types) and may be persisted in a database if needed. The persistent representation form is the master copy of a BO and is hosted in a unique application or service (the object's owner), which is responsible for the BO's life cycle. All other types of representation (and even other persistent representations) are object copies related to the object owner.

OAGIS is mainly used for standardized communication between different tools. Our BO model, however, and especially its implementation, has a much broader scope: as stated above, it is also used for Java communication and for persistence. This implies many difficulties in implementing the model, e.g. automatic marshaling and unmarshalling between Java and OAGIS-XML, as well as consistent strategies for lazy fetching and cascading of create, update and delete operations on such complex data structures. The BO Instance Repository implements all this consistently.

Since the OAGIS overlay schema is very complex and uses all XSD features, we automatically derive a simplified XML schema from the standard schema. This simpler schema can be used alternatively to construct and validate XML instances of BOs, without any restrictions. However, it only uses a simple subset of XSD, e.g. no inheritance, substitution groups, choice and others, and therefore even rather simple tools can use it.

2.3 Domain Model Development Methodology

Developing a data model containing all business objects needed in the considered domain and identifying the existing connections is a very complex task. In accordance to the basic understanding in systems and software engineering, the following phase-based and iterative approach proved to be suitable and was applied for BO4L.

Stage 1: **Problem and Requirements Definition**. In the course of this stage, typical workshops with experts of the considered domain should be held. Supported by the use of creative methods such as concept card and brainstorming, the relevant requirements should be identified. For details concerning requirements of the BO4L domain model, e.g. see [12].

Stage 2: **Analysis**. The goal of this stage is to identify every needed BO within the considered domain. As predominant standards developing organizations like Open Application Group (OAGi) or GS1 have shown with their business document standards OAGIS [15] and GS1 XML [22], the best way to identify business objects is based on scenarios. However, experience within the team of problem analysts has shown that, in order to facilitate BO identification, scenarios should be described on a much more detailed level than e.g. OAGi and GS1 use within their scenarios. This stems from the fact that not only business-to-business (B2B) communication should be facilitated, but also communication between business process applications with one application being just a small fraction of today's range of functions of business applications, i.e. receiving or inventory control.

Stage 3: **Design**. Within this stage, identified BOs are designed. In this context, design means firstly the modelling of the business object's associations and secondly the definition of all its attributes. Revealing associations is thematically very closely linked to Stage 2 and may lead to frequent returns and reassessments of its findings. Therefore, it seems to be more efficient in terms of design efforts to first create associations and then to design the object itself in more detail.

Certainly, the object´s attributes are closely related to the scenarios. Again, scenarios play a key role in this stage. However, even though they are inevitable, they should be enhanced with information provided by existing standards in order to gain completeness of attributes—as far as possible. In case of the BO4L development, well-established business document standards (OAGIS [15]; GS1 XML [22]; EDIFACT [20]; openTRANS [23]) were chosen as references. The working group constructed a model that can be regarded as an evolution of these standards. In this way, it was assured that the resulting business-IT gap is minimal. Finally a first UML model results.

Stage 4: **Maintenance**. It is clear that no model is comprehensive and faultless right from the start–no matter how smart and thoroughly the design process is. Maintenance is therefore a key element of every model´s life cycle. The way this is managed by model governance will be explained within Sect. 2.5.

2.4 Technical Model Development Methodology

The development of the technical model is also a multi-stage process, consisting of the following steps:

Stage 1: **Selecting a Basic Standard**. For yielding acceptance of the Logistics BO model, it has to be based on accepted standards. Several transaction standards were evaluated, as described in Sect. 3.1, and OAGIS was selected.

Stage 2: **Producing a Light OAGIS Overlay**. The outcome of the functional analysis is materialized as an XML schema. The base standard OAGIS is adapted in a standard-conform way, so that it fulfils all our requirements. An extended subset of OAGIS results, where also the extension consists of standard components. This construction process has the following steps, as depicted by Fig. 8.

- *Removing nouns*: All OAGIS nouns are deleted, which are currently not needed, i.e., which do not correspond to a supported BO. These BOs may be reactivated later, as needed.
- *Construction of the Overlay* by standard means: attributes needed, but missing in OAGIS are added mainly by sub-classing the OAGIS noun. There is an OAGi guideline document on how to do this, see [24] and its addendum [25]. Moreover, even new nouns with all their needed attributes are added as needed.
- *Sample instances*: for all components and all BO small sample instances are constructed bottom-up, containing only the supported attributes. This process is guided by a structural containment graph, constructed beforehand using algorithm developed earlier. The samples need not contain meaningful data, but with meaningful data they constitute an important artefact for users of the technical BO model.
- *Lightening the Overlay* according to the small sample instances: by using the free tool *XMLHelpLine*, the XSDs of step 2 are reduced to contain only those definitions available in the sample instances from step 3. This is a fully automatic process. Optionally, the tool can also do a flattening, reducing the number of files resulting.

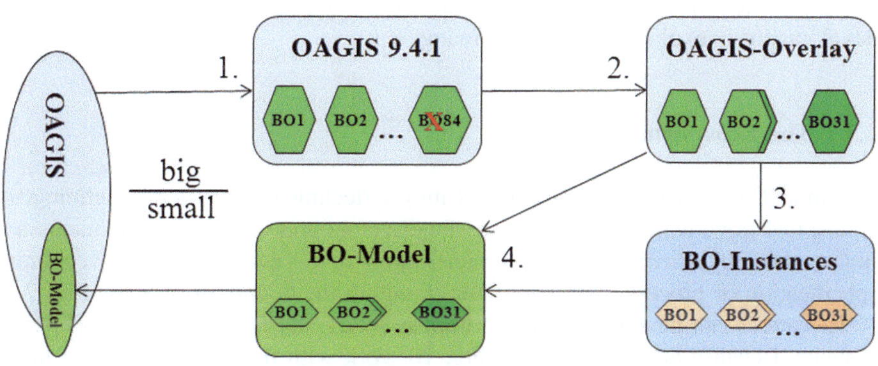

Fig. 8 Construction of the technical business object model as a light OAGIS overlay

Stage 3: **Constructing a better Domain Model**. After these steps we had two model representations: the initial UML model from the functional analysis and the OAGIS overlay. Our goal was to use the UML model also for generating other technical artefacts, namely the Java entity model. In Sect. 3.1 we describe that the construction of the Java entity model is a manual engineering step. It could not be derived from the technical OAGIS representation, but the UML model has been redesigned to get all model representations in harmony as much as possible, and to use automated steps for the development.

Stage 4: **Artefact Generation**. Many of the BO model artefacts are constructed automatically: The technical model is based on an automatically lightened OAGIS overlay, complemented by a Java entity model generated from the domain UML model, which is therefore a business-oriented Java entity model. From the Java model, a database model is generated using JPA, used by the BO Instance Repository. Moreover, BOD XSD files and WSDL files are generated based on templates and configuration files.

2.5 Model Governance

The set of BOs and their associations form a consistent data model of the logistics domain. As stated above, it is impossible to represent this complex industry with all its sub-disciplines completely. These circumstances led to a scenario-based approach for defining BOs (see Sect. 2.3).

Furthermore, practical experience proved that such domain models are subject to change as frequently as the domain itself. Hence, a fundamental assumption is that the BO model is experiencing permanent development. Dealing with the change becomes a crucial challenge.

Model governance is mainly driven by the functional model. Besides the functional model, several other aspects are affected by changes, i.e. documentary reports and technical models. OAGi describes its development methodology in [26]. For the technical XSD part, this has largely been adopted. The developed process model for the evolution of the functional model as driving force for further development is explained in the following.

2.5.1 The Ecosystem

The model governance is incorporated into a "technical ecosystem", defining the concept of governance, identifying its inner actors and technical components and linking those by activities. Several policies apply to this ecosystem like a versioning specification or process instructions for changing the definition of a BO Fig. 9.

Resources can be subdivided into human actors, the BO model, and (IT) components. Because all roles listed in Fig. 10 work with those BOs, they are incorporated into the ecosystem as users (stakeholders). Two more groups of actors have

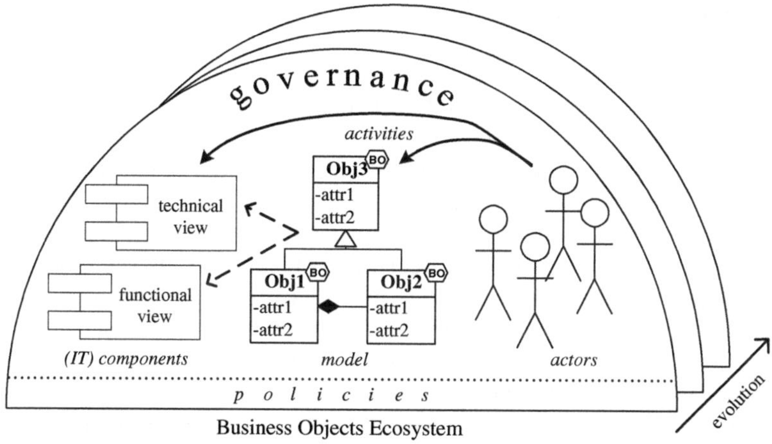

Fig. 9 Business Objects Ecosystem

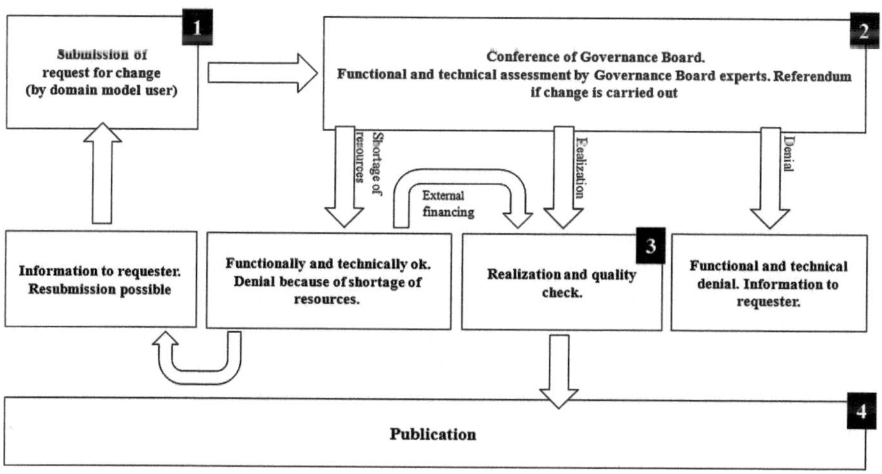

Fig. 10 Governance process of the domain model

been identified: people who actively develop the model (developers) and a board deciding in case of doubt or conflicts. Without such a board, the process of development would certainly become ineffective due to the heterogeneous participants and their conflicting intentions. The BO model affects several (IT) components, which in most cases are artefacts of the overall Logistics Mall. The functional and technical views on the model are of great significance in this context because of their mutual dependencies.

Interactions among resources are represented by combining specific resources with an activity to form a use case.

2.5.2 Evolution of the Domain Model

Everything in the ecosystem will change over time. These changes occur on two different levels: Modifications of the BO model or their dependent components are considered substantial (content level). Controlling these is the main purpose of the ecosystem. However, in case a policy changes or when a new activity is added, change takes place on a meta level.

The ecosystem can control itself by corresponding (meta) policies, e.g., a policy on modifying content-level policies. Because the focus certainly is on the evolution of the BO model, the developed change process is presented here (see Fig. 10).

This BO model maintenance process has the following steps:

Phase 1: Change Request. The process already starts with the necessity to change the model. This can be expressed by every arbitrary combination of the following actions involving one or more BOs: Adding a new attribute/association, modifying an existing attribute/association, deleting an existing attribute/association. All developers within the ecosystem (see above) are eligible for submitting their suggestions to the board.

Phase 2: Assessment of Requests. The Governance Board checks the technical and functional compatibility with the existing BO model. Besides compatibility, the decision on the approval can be influenced by strategic and economic aspects. Changes could be denied because of functional or technical weaknesses, strategic aspects, or economic aspect, e.g. shortage of implementation resources. This could then lead to adjournment and resubmission. In both cases (denial and adjournment) the requesters are informed.

Phase 3: Realization and Quality Assurance. In case of approval, the changes are implemented into the BO model. The process of implementation is again subject to policies, containing rules and maxims, set up for a consistent and high-quality result, independent of the individual who performs the changes. The main result of this phase is a new BO model release. Due to the dependencies, new (versions of the) components are produced as well. During this phase, the board checks the compliance of the implementation regarding the policies of phase 2. In the case when discrepancies are found, phase 2 is performed again. When the new model and the components are finally approved, they can be published.

Phase 4: Publication. A new release of the BO model is made available to the users of the ecosystem. However, not every implemented and approved change request will result in a new release. Changes performed within a certain period are combined into a single release. There are a couple of important aspects to be considered for the process of evolution not mentioned here. All of these are covered by dedicated policies. Examples are versioning and the lifecycle of a release or version. They are introduced in the subsequent section.

Fig. 11 Levels of versioning

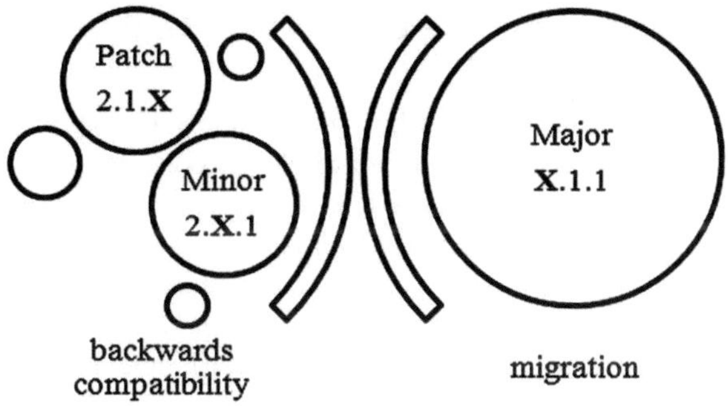

2.5.3 Versioning Concept

Versioning is a big issue within the topic of Model Governance. On the one hand, versioning should facilitate evolution. On the other hand, it must guarantee stability and reliability, as these are crucial for its application and success in practice. Therefore, any model governance should come with an adequate concept of model versioning. Especially, the aspects of backwards and forwards compatibility need detailed attention. Other aspects are version lifetime and the number of parallel valid versions, since these effects in a large part costs for maintaining of several old versions.

Figure 11 shows three levels of versioning concepts considering these constraints. Revisions on the lowest level (patch level) and the medium level (minor release) are backwards compatible. While changes on patch level only include smaller optimizations and bug fixing, minor releases include adding of new BO and BO attributes and technical changes. Major releases allow substantial changes of the domain model, e.g. changes in structure and deleting attributes and objects.

As major releases are not backwards compatible, migration efforts on App-level have to be carried out, i.e., all apps using these BOs in their new mayor version have to be modified. Thus, major releases are produced once a year and a lifespan of five years seem to be appropriate (see Fig. 12).

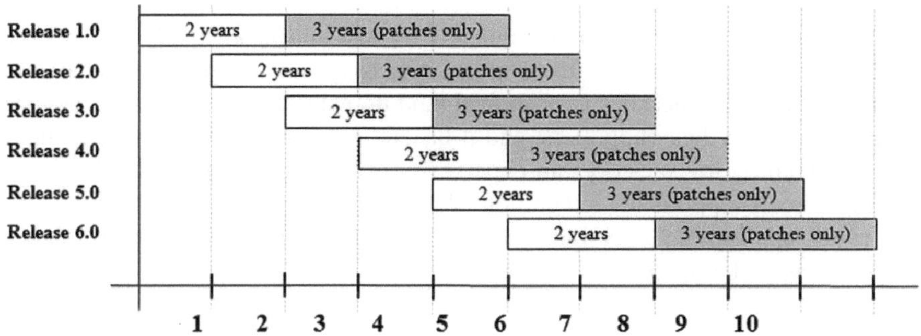

Fig. 12 Lifespan of domain model versions

3 Narrowing the Business-IT Gap Using Business Objects

This section describes how the business-IT gap and the two horizontal gaps mentioned in the introduction are closed concretely, i.e. how the vision initially sketched by Fig. 1 is realized.

For lowering the horizontal gaps a common model is used, this is the domain BO model for the business-business gap and its canonical[1] technical representations for the IT-IT gaps. Thus, the BO model is a common language with two dimensions: business and IT. The business dialect is understood by business people and the IT dialects are understood by different layers of IT systems.

In Sect. 3.1 we discuss the horizontal IT aspect, namely canonical application communication, orchestrated by protocols. The horizontal domain aspect was already discussed in Sect. 2.

In Sect. 3.2 the business-IT gap is closed on data level by mapping the business view of BOs to its IT view. Then Sect. 3.3 closes the business-IT gap on activity level, where domain tasks are mapped to IT system APIs if they can be automated-manual tasks remain. Section 3.4 finally connects all aspect on the business process level and summarizes the argumentation.

3.1 Horizontal IT Gap: Canonical Application Communication

Logistics services or applications like a warehouse management system, a transport management system or an ERP (enterprise resource planning) systems and even smaller logistics modules and services, which we call apps, are all seldom used stand-alone—they have to cooperate, e.g. by exchanging messages. Interoperation is needed for the different applications of a customer all available via the Logistics Mall (i.e., in the Cloud) and for the apps of his cooperation partners in the Mall Cloud, but also between applications in the Cloud and applications the customer already uses and still wants to use on premise. This requires a common data model to reduce integration efforts.

The Logistics BO model is canonical [27] and *independent* of any software product or software provider. Even more, it is based on accepted technical *standards* constructed in decades by technical (and domain) experts from the standards organizations and their supporters, and it has an abstract domain view constructed and discussed by logistics domain experts from two Logistics project clusters.

Therefore, it supports the integration of different applications in an optimal way: For every application, only one business model adapter has to be developed, independent of the number of its communication partners, as shown in Fig. 13 on

[1] canonical means conforming to accepted principles or standard practice, i.e. being harmonized, agreed on, minimal, standardized, reusable, but also extensible [27].

Fig. 13 Canonical communication via standard business objects

the right side. This contrasts with the left side, where for each communication partner a different interface has to be provided and maintained, and different languages are used and many mappings are needed, since same things are named and structured differently.

3.1.1 Selecting a Base Standard

Widely accepted e-business standards are central for interoperation of applications, and they are used on top of technical interoperation standards. Not only the business objects exchanged need to be standardized to ease communication with different kinds and instances of other applications, the protocol steps have to be considered as well.

Today, many e-business standards exist, as described in Sect. 5. We compared and evaluated different transaction standards and selected OAGIS [15], the Open Application Group Integration Specification. The reasons are:

- it is a powerful, adaptable and extensible horizontal standard, based on Core Component Technical Specification CCTS [13] and many other standards
- it is mature (first version in 1996—pre-XML, we used version 9.5.1) and in real use (broad industrial support in many countries)
- a logistics module and many logistics-related definitions already exist (e.g. container, shipment, item, pick list), only few extensions are needed for logistics
- it is technology-neutral, but includes Web service definitions, and there are OAGI working groups on Cloud, Mobile and BPMN support.
- UN/EDIFACT [20] formats (in widespread use) are convertible to OAGIS.

3.1.2 Benefits of Using Canonical Standardized Communication

The benefit of using such a widely accepted standard for canonical communication is that the format of most messages ever needed is already defined based on the long experience of a large expert group. Only few definitions have to be added, based on

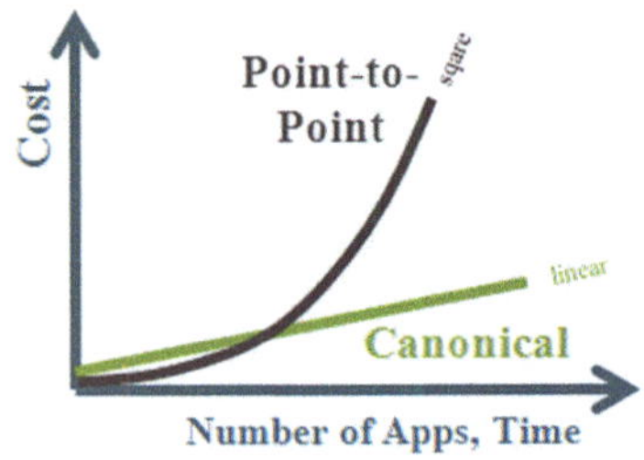

Fig. 14 Cost benefits of canonical communication

the standard's extension mechanism. These definitions are in turn based on CCTS [13] and its Core Component Library CCL, a mature reservoir of components used as semantically described building blocks in OAGIS and several other standards. Thanks to its many optional attributes, even simple messages can be exchanged conformant to the standard. If special information is needed, its position and format is already defined, thus minimizing data mapping needs.

This enables stable, standards-based interoperation interfaces of applications or services, initially implemented as needed (filling only needed data elements) and subsequently completed as needed in a downwards-compatible way. The same standardized interface can be used for many different partners, and additionally EDI converters can be provided: as OAGIS itself is based on UN/CEFACT standards, transformation to its grandpa UN/EDIFACT still widely in use in logistics is possible and supported.

As depicted by Fig. 14, when using a canonical model, the cost for integrating a small number of apps is initially higher, but only increases linearly with the number of interoperating apps. Especially when modifications of apps occur over time, the canonical communication is beneficial.

3.1.3 Technical Communication via BODs

OAGIS defines several BODs (business object documents) exchanged between applications. Each BOD is composed of an *Application Area* (the document header) and a *DataArea*, consisting of a verb and one or more nouns. The verb defines the action to be performed on the nouns or data, which represent the main BOs, visible in the application or service interface.

For each noun, there is a service having methods representing meaningful verbs of that noun. This service can be implemented by various technologies, e.g. RMI (Java Remote Method Invocation), REST or as a Web Service. BODs are used as message formats for these methods, regardless of the technology used. OAGIS itself comes with sample WSDL definitions (the OAGIS services), but we added and implemented services for RMI and REST.

Each application or service running in or outside of the Logistics Mall requires a Mall interface consisting of its required OAGIS services in a technology. This interface has to be implemented towards the enterprise service bus (ESB) used in the Mall or towards the BO Repository. It can then communicate with other application also implementing such standard interfaces. Besides this standard communication, peer-to-peer-communication using other formats is still allowed to smooth the integration curve, e.g. by using EDI converters.

For efficiency reasons, the Logistics Mall internally often uses call by reference. For this purpose, we defined *MiniBODs*, which are normal BOD, but mainly contain identifiers of BOs instead of copies of BOs. These MiniBODs uniquely reference BOs stored in the BO Repository.

3.1.4 Communication Protocols

To orchestrate the application or app communication, different scenarios and protocols can be realized based on the asynchronous exchange of BODs, as depicted by Fig. 15, an UML sequence diagram. Here, some customer orders goods from a supplier, and for the shipment, a logistics carrier is used. The process starts by

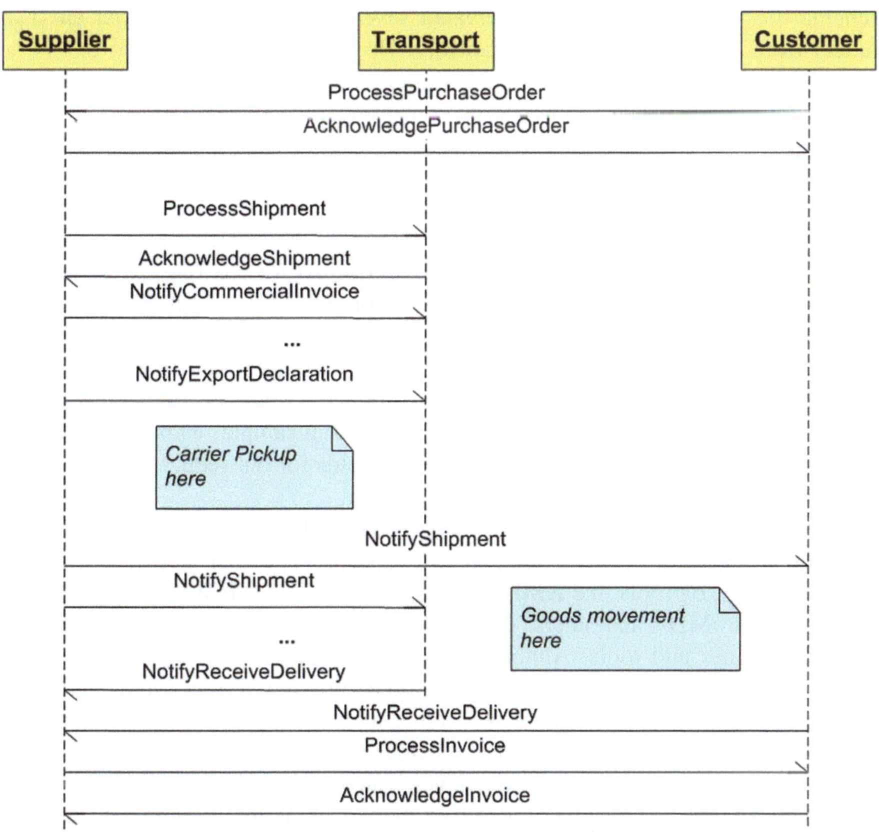

Fig. 15 OAGIS logistics scenario no 59-Extract of [15]

sending a *ProcessPurchaseOrder* BOD message from customer IT to supplier IT, and the latter acknowledges the order using a BOD. Communication proceeds as needed, until finally the invoice is sent to the customer IT and acknowledged. Such diagrams will soon be available in BPMN from OAGi.

As stated above, these BODs are message formats in OAGIS services, which constitute the OAGIS application interface services of involved parties. Such protocols are often hard-coded in the applications, where receiving a BOD will eventually result in sending the answer BOD, either synchronously or asynchronously.

A more flexible and more abstract orchestration of services can be defined using business process models, as described in Sects. 3.3 and 3.4. As a preparation, we need a mapping of the domain business objects model to its technical representations.

3.2 *Vertical Data Gap: Mapping the Domain Model to the Technical Model*

As OAGIS is our underlying base standard and OAGIS is defined by a set of XML schema files (XSDs), we firstly evaluated several methods to construct an UML model from these XSD files automatically. However, it turns out that this is not feasible to construct a domain model—a good abstraction is a manual engineering step, and the mapping has to be stored somehow.

3.2.1 Automatic Mapping Approaches

OAGi itself used the free Eclipse plugin *HyperModel* for creating UML diagrams for their definitions. HyperModel is able to import XSD files and to display them as UML diagrams, and it supports some operations on the diagrams, e.g. merging inheritance. However, the resulting UML is very complex and not at all understandable by domain experts. The UML diagram for a single business object would fill a wall if expanded in all detail. See Fig. 16 for a sample, namely a *PurchaseOrder* (still with unexpanded parts). This even holds for our reduced overlay, since HyperModel displays all the details in any depth.

The next approach was to generate Java from XSD, and then to generate UML from Java. There are several tools for both steps. The first step is critical and supported, e.g., by tools like XMLBeans, JiBX, JAXB, CXF WSDL2Java, Castor, Smooks, AndroMDA and others. The interested reader may easily find references to these techniques and tools. Although some tools had difficulties with some constructs used in the OAGIS XSDs, it worked, but the results were again very complex-the resulting Java model was rather unusable. Also for the next step, the mapping of Java entities to database schemas to implement BO persistence, different OR-mapping techniques were investigated (DataNucleus, JDO, JPA, etc.) and even NoSQL approaches, but all of these techniques require a Java model simpler than generated.

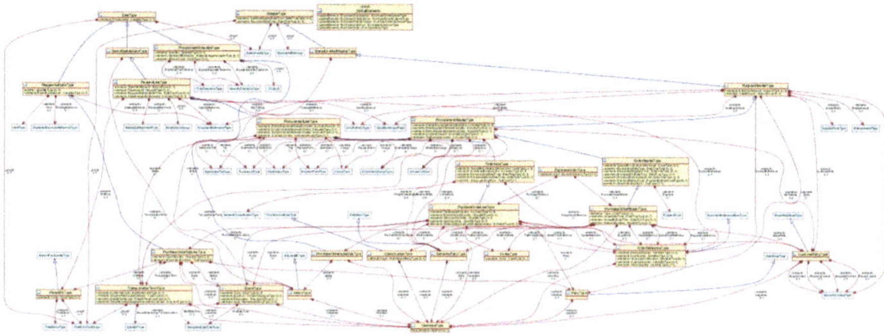

Fig. 16 Inappropriate domain model: Part of the OAGIS purchaseOrder noun

3.2.2 Engineering the Mapping

Therefore, we understood that constructing a good domain model is a manual sophisticated engineering task, requiring excellent domain knowledge, even when being directed by a highly detailed and elaborated technical model like OAGIS. Sophisticated abstraction is needed to hide the complexity (which is needed for tool integration) from the domain expert, and meaningful abstraction cannot be generated. On the other hand, the abstractions constructed should be compatible to the original model, since a simple mapping from the domain model to the technical model is essential for our approach, as described in Sect. 3.

Figure 17 illustrates the construction of the logistics domain model as an abstraction of the subset used from the technical model: unused attributes are ignored and long paths are compressed. New nodes are sometimes inserted to regroup the result and increase comprehensibility. The abstraction was developed by several domain experts; therefore, it is understandable for logistics domain users having no technical knowledge. The result for the BO *PurchaseOrder* is shown by Fig. 4 in Sect. 2.2. This simplified model has several good properties:

- It is simple enough to be understood by domain users
- It is simple enough for further generation steps, e.g. for the generation of persistence support
- It can be extended by methods, i.e. adding behaviour to the structure, at least CRUD methods (Create, Read, Update, and Delete) with some sophisticated aspects like CRUD on collections and managing inverse relationships.

3.2.3 Maintaining the Mapping

Each attribute in the UML model is mapped to the complex standard OAGIS XML path, stored in the BO Metadata Repository (see Sect. 4.2). The resulting benefits of this construction are as follows:

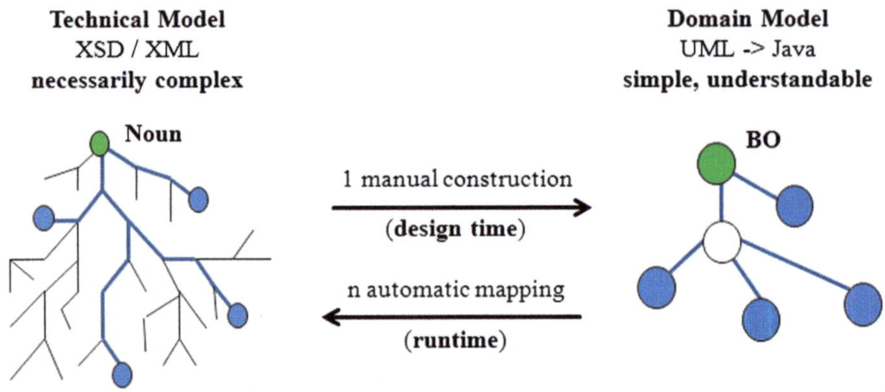

Fig. 17 Mapping the domain model to the technical model

- The domain expert has a simple and understandable view of an underlying complex model and can use it e.g. in process modelling.
- The IT expert has a detailed technical model and even several technical representations, capable for standard-based tool integration and communication with other tools.
- Both models are tightly coupled and automatic translation is possible
- The business-IT gap is minimized, since the domain model BOs are pre-implemented in Java and as standard OAGIS-based XML schemas. The Java mapping is 1:1 UML to Java generation, while the OAGIS mapping is 1:1 stored paths. Domain artefacts are pre-linked with different technical artefacts, as described in Sect. 3.3.

3.3 Vertical Behaviour Gap: Pre-implemented Activities for Process Models

The business-IT gap not only exists for data, but also for handling and manipulating data, i.e., for business tasks performed by IT systems, if the tasks can be automated. The business tasks deal with the business view of data, while the IT systems need technical representations. Process models can be used to orchestrate business tasks, not only automatable tasks but also manual tasks, executed with the help of software or purely manually with and on logistics BOs.

3.3.1 Business Process Models

In times of international competition and tight profit margins, efficient business process management including partner interactions is essential for logistics companies, especially for SMEs. Since prices are dictated by the market, only the internal optimization of process models can help setting some unique selling points: Only individual process models, not the standard ones, can make a difference in the market: the unique selling points.

Business demands change quickly, especially in logistics. The business processes of a company should be quickly adaptable to new requirements. However, often a quick adaptation is not feasible, as the implementation of new logistics process models, designed by logistics experts, also requires additional IT support that takes time to be brought in place.

To speed-up process design and implementation, the Logistics Process Designer LPD [28], as part of the Logistics Mall, enables quick process adaptations without the need for in-depth IT expertise and enables the deployment of these process models merely by a click of a button. It is based on standard BPMN 2.0 [29], generated from a simpler domain-oriented process model, and used for process execution in the Cloud as well: the business users can focus on their business processes–they do not need to care about the process engine and IT concerns.

Business processes are modelled by domain users, not by IT experts; therefore, process models focus on control flow. However, data flow can also be modelled: in domain view, activities have inputs and outputs, both described by some BOs from the domain model. The inputs of an activity have to be produced by some predecessor activity.

3.3.2 Technical View on Processes

Technically, each process instance (being a started process model) has its local process context, and each activity in the process instance can access and write data in its process context. Activities are implemented by API methods of IT systems or by special services or apps.

To minimize the business-IT gap in process modelling and process execution means to minimize process implementation efforts, since normally process implementation needs IT experts. Here we use building blocks on domain level as depicted by Fig. 18, which are pre-implemented by IT system methods and used as a modelling unit.

These building blocks have to specify how the business activity communicates with its IT implementation. This is eased by pre-implemented BOs used in the domain view and pre-mapped to the technical view. They close the business-IT gap for data and for behaviour.

Such a block mainly consists of additional metadata for IT systems and apps offered via the Logistics Mall. The metadata describes the names, inputs, outputs

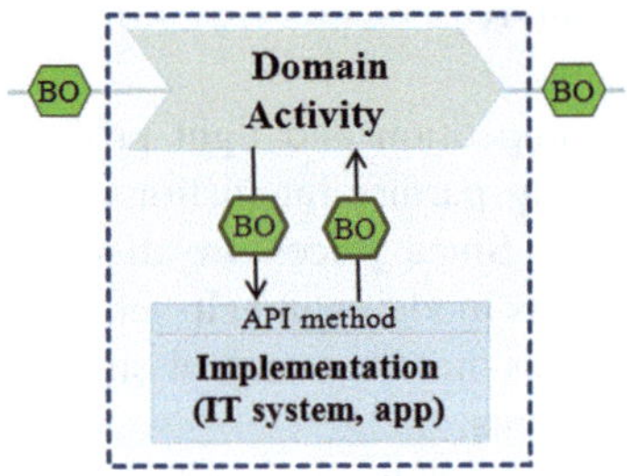

Fig. 18 Pre-Implemented activities as building blocks on domain level

and behaviour of all activities which can be performed using the offered system and is used to generate BPMN 2.0 XML task descriptions.

Since our Logistics BOs are rather complex and may become very big, we relieve the process instances (or better: the process engine's process contexts) from handling them–the process engine only stores IDs (i.e. references) of BO instances which are stored in the BO Instance Repository of the process owning party. In this way, the IDs exchanged between process instances and apps coordinate their data transfer with the BO Instance Repository, while the processes organize the control flow between apps.

3.4 Narrowed Gaps: Paradigm Shifting in Process Modelling

The canonical BO model is central for the orchestration of Logistics Mall products (apps) by using business process models, designed and deployed by business people (with minimal IT support). The mechanism allows phase inversion, since it is based on pre-implementations of two kinds of modules:

- *Behaviour modules*: Activities for offered Logistics Mall products using BO-based interfaces, as described in Sect. 3.3
- *Data modules*: standardized product-independent BOs, as described in Sects. 3.2 and 2.2

3.4.1 Phase Inversion

Most BPM suites have different tools for process modelling by domain experts (phase 1 in Fig. 19, modelling a business process using activities *A*1, *A*2, etc.) and for the later implementation of these process models by IT personal (phase 2, implementing each *Ai* by some implementation *Ii*, e.g. by a service call). However, in the Logistics Mall we deploy and execute modelled business processes merely by a click of a button, not involving IT knowledge or IT personal.

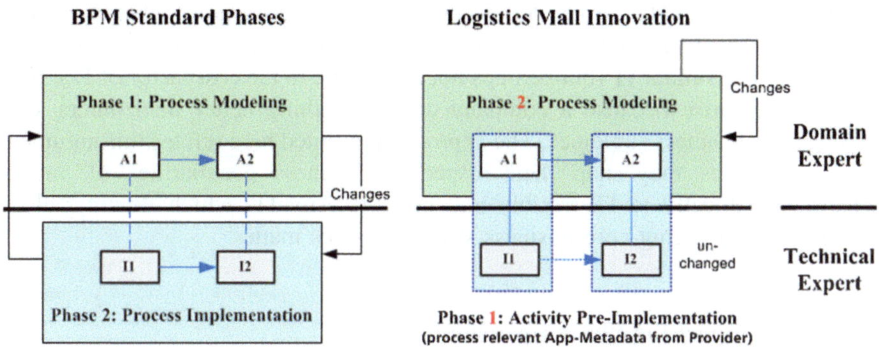

Fig. 19 Phase inversion-conventional approach versus logistics mall approach

This is only possible, if we model with already implemented building blocks with standard interfaces using standardized data formats, since implementation after modelling should be minimized. In this case, we have a phase inversion, also depicted by Fig. 19: the application or service provider (i.e. the provider of *I1, I2*, etc.) not only provides normal metadata, but also process model-related metadata for his offer (phase 1). This serves as implementations of activities used in process models, since from this metadata, activity descriptions in BPMN 2.0 XML can be generated. Process modelling and modifications of process models are then the *next* steps. They have no further influence on activity implementation. The same activity implementation can be used in different process models of the same or of different tenants, and in any new versions of these process models as well. The modelled process can be changed and is always directly executable, since it uses pre-implemented building blocks using standardized pre-implemented data models in its interfaces.

This principle is sometimes called "What You *Can* Model Is What You Execute" in analogy to the WYSIWYG principle of text editors. Here the word *Can* is important, since of course this modelling approach has some restrictions, in analogy to Lego© construction sets.

BPMN 2.0 optimally supports this idea, since it uses the same notation for process modelling and for process model deployment and execution. The latter has additional attributes set in its standardized exchange format BPMN 2.0 XML. Here, the task description with all implementation-relevant attributes of the process-relevant service metadata is inserted for the modelled tasks.

Moreover, the idea is also practical in the context of the Logistics Mall, since the scope of its process models is limited to the logistics domain, limited in complexity (if complex logic is seen in the applications) and even stronger limited to only orchestrate concrete applications and services offered via the Logistics Mall or made accessible to the Logistics Mall. Basic services (such as a mail service) will be offered free, to get a critical mass of building blocks. In [30] it is described how top-down modelling can be achieved.

Nicholas Carr [31] wrote "IT Doesn't Matter": companies cannot achieve competitive advantage only by IT, since IT is no limited resource and other companies may use similar IT (but perhaps not SMEs due to the cost factor of IT). Such benefits only arise because a company does something better than others, concerning their business strategies. The approach presented here refutes this argument, since companies may map their respective business peculiarities of various departments directly to IT flexibly and quickly. Any IT, which flexibly allows realizing and changing core business strategies, does matter.

3.4.2 BO Support for Phase Inversion

The logistics business object model BO4L follows the same principle in its two views. There is an abstract domain view defined by UML diagrams, which hides the complex details of the underlying OAGIS Logistics Overlay. The abstract model was constructed based on the technical standard (again some kind of phase inversion) by several logistics experts of Fraunhofer IML and in conjunction with other Leading-Edge Cluster logistics projects, to assert both the refinement relation of the two models and at the same time the understandability by domain experts.

While the technical model is in English only, both the domain model and the application or service metadata and process metadata are multilingual and commented (descriptions can be displayed in the process modeller in a selected language). The domain views are only used during process modelling or service provision, while the technical models are used at runtime.

We call this the double barbell principle [30], since for both, behaviour (the process model) as well as data structure (the business objects), the corresponding domain view and technical views are closely related in advance like a barbell (compare Fig. 20). The technical views cannot be constructed as a later refinement (of arbitrary BOs or activities), since IT personal should not be involved. Instead, the barbells are constructed only once by the providers of applications (and the BO

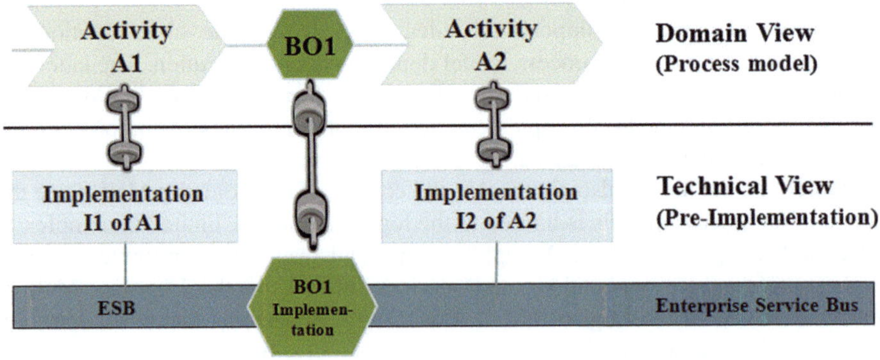

Fig. 20 Barbells pre-connecting domain view and technical view of data and activity

model governance team) and then reused in any process models. The standard BO model is important to avoid complex data mappings, as the standard activity model is to avoid complex implementation. Both are done only once, not for any new process model or model changes. This is especially important in a Cloud environment, since Web interfaces for the implementation tasks are difficult to develop.

Compared to Fig. 1 in Sect. 2.1 the GUIs of IT applications or apps are replaced by activities, which may be manual activities implemented by the GUIs or automatic activities implemented by the IT systems or apps.

3.4.3 Putting Everything Together

Figure 21 depicts the resulting architecture with all gap reductions: The upper part is business: process models are used to describe business processes handling business data, indicated by dashed lines in process models. Data flow in process models can be specified, but in principle is bus-like—every BO produced previously in the running process can be accessed (from the process context). The process models together with a common data model, the Logistics BO model, bridge possible business-business gaps, e.g. between cooperating partners.

The bottom part is IT: different apps (i.e., implementations of activities) communicate via technical BOs in Java or OAGIS XML representation, and here we use the BO Instance Repository as a counterpart to the process contexts of all process instances maintained by the process engine. The process instances only

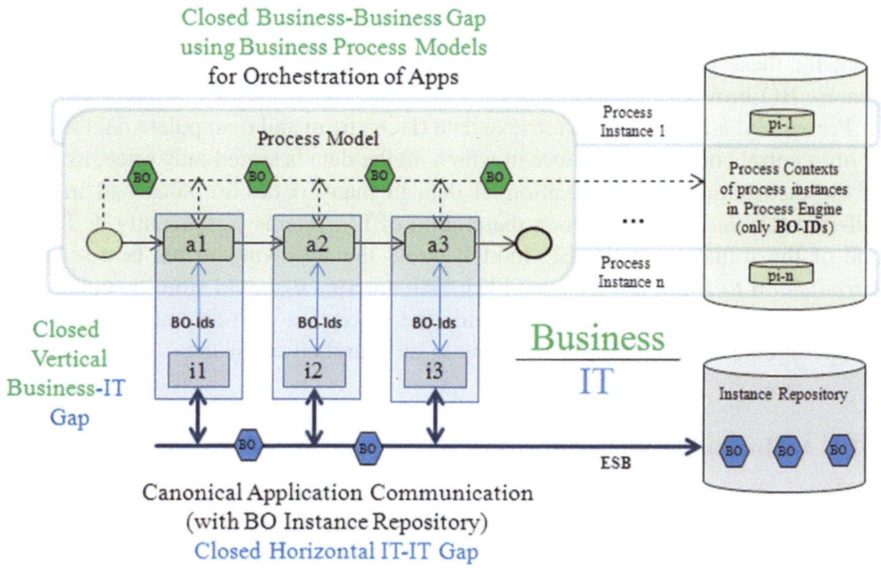

Fig. 21 Narrowing vertical and horizontal gaps

handle BO-IDs, which are used by the orchestrated apps to access and exchange data. This closes the IT-IT-gap.

The most important part is closing the vertical business-IT gaps for data as well as for each activity. This is done by pre-implemented BOs and pre-implemented activities generated from the Logistics Mall offers and operating on BOs. These connected building blocks composed of a business item and an IT item are flexibly orchestrated by business process models, resulting in a "What You *can* Model is What you Execute" approach.

4 Software Components for Seamless Integration

The business-IT gap is narrowed even more by proving supporting components that ease BO model handling and seamless integration. It is important for the approach presented in the previous sections to get support for BOs during design time (e.g. when designing business processes or when registering Logistics Mall app offers) as well as during runtime (i.e., when executing business processes and apps). For this purpose, we offer the BO Metadata Repository and the BO Instance Repository.

4.1 BO Instance Repository

Having a common data model (i.e., the BO model) used by all tenants and all apps is a central idea of the Logistics Mall. Then, providing a logically common data store for these BOs is a logically next step. This BO Repository also provides a generic BO browser.

Preferably, all apps should retrieve data (BOs) from and manipulate data residing in one central common data store in which all the data is stored only once, to get rid of the distribution and duplication of data in many data base silos and in many different formats, which hiders a maturation of IT systems significantly. C.J. Date, one of the founders of the ER model, wrote the following in his best-seller *An Introduction to Database Systems* [32]: "An enterprise should store its operational data in an integrated database to provide the enterprise with centralized control of its operational data". At least there has to be a unified consistent view of data.

4.1.1 A Motivating Scenario

Logistics applications like an ERP system, a WMS or a CRM system are seldom used stand-alone-they have to communicate. i.e., exchange master data as well as transactional data, not only the systems of one tenant, but also those of his cooperation partners, being in or outside of the Mall. This is depicted by Fig. 22: three

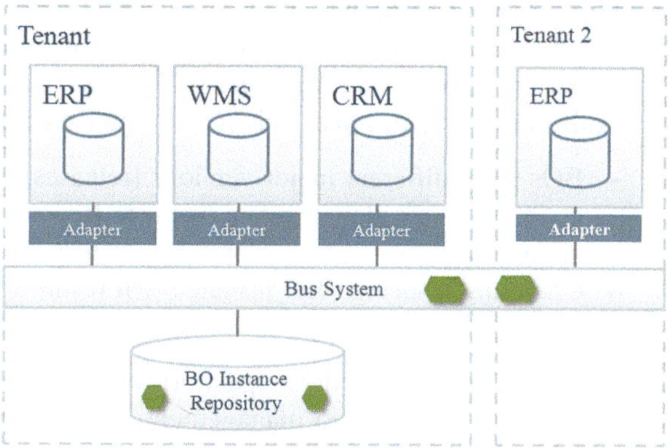

Fig. 22 Scenario motivating the BO instance repository

logistics applications are integrated as offers of the Logistics Mall and rented by a tenant, and they cooperate with an ERP of another tenant.

The application providers make their offers Mall—conformant–for existing systems, they write adapters for standardized BO communication using the Mall bus system. They have two choices: exchange BOs over the bus, or use the BO Instance Repository as a common data store, for all own apps or even for apps of communication partners. The latter choices have additional advantages, especially for newly developed apps.

4.1.2 Objectives and Requirements

The objectives of the BO Instance Repository are as follows:

- **Supporting App Integration**: The BO Instance Repository can easily be integrated into existing logistics applications or Apps. Alternatively, the data tier of apps is given by the Instance Repository, which pre-implements the standardized BO schema.
- **Data Consistency**: All Apps have access to up-to-date and consistent business object instances.
- **Security**: All Apps can only access or modify data for which they have corresponding rights.
- **Saving Costs**: The integration of apps is eased, data synchronization is automatic, and any unnecessary data conversion is avoided.
- **Cloud-Based**: the Instance Repository must fit seamlessly into the architecture of the Logistics Mall cloud.
- **Distributed**: to avoid possible bottlenecks of a centralized data source, the components of the repository architecture can be distributed to different nodes.
- **Model-Completeness**: each Instance Repository is able to store instances of all supported BO types (i.e., it covers the complete BO model). This supports integrity and consistency.

- **Multi-Tenant Support**: All Logistics Mall customers have to be cleanly separated. Here we use private Instance Repositories for each tenant. This eases migration, security, backup, and scalability.
- **Multi-Version Support**: This has a type aspect and an instance aspect:
  - BOs from different major versions (releases, see Sect. 2.5) of the BO model are always stored in separate Instance Repositories, while patches and minor releases use the same Instance Repository. This eases data migration.
  - Changing a specific BO instance will result in creating a new version of this instance if the BO type is versioned—otherwise this change is only logged.

Normally, such central data repositories are used for master data only. In contrast to conventional master data management, we may not only store master data and structural data in this repository, but also dynamic data and transactional data. The following benefits result: BOs of any kind can be exchanged by reference (not necessarily but copy), all (access and) changes are automatically logged, and the implementation efforts are minimized.

4.1.3 Architecture

The BO Instance Repository follows a multi-layered architecture as depicted by Fig. 23: The central part is the BO Entity Model, which is synchronized with the BO model in UML (the Java code is generated and can be reverse engineered with the UML tool).

The UML model even contains annotations to configure the database access layer (JPA annotations). This layer is independent of the storage technique used: although relational database systems are a first choice to store the data physically, JPA even supports cloud databases and other techniques.

On top of the entity model there is a small business logic layer supporting the creation, update and deletion of BOs as well as reading and different methods of

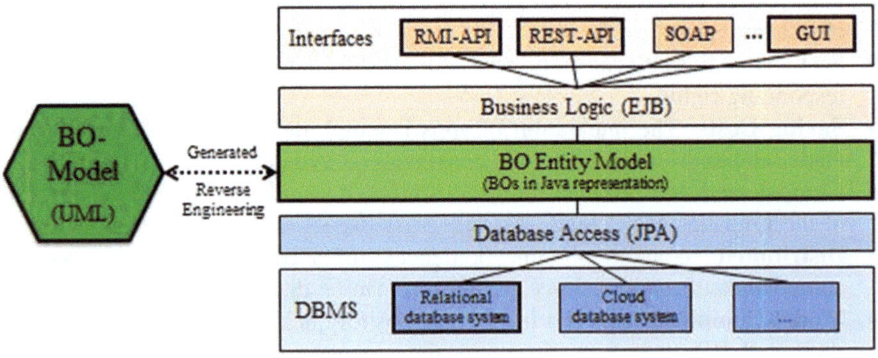

Fig. 23 Architecture of the BO instance repository

searching BO. This functionality can be made available via different interfaces (APIs, application programming interfaces), e.g. a Java RMI interface, a REST interface or SOAP Web services, and additionally a BO browser (GUI), which allows to inspect every detail of the content in the BO repository and to search for information. BO manipulation support via the BO browser is planned.

4.2 BO Metadata Repository

To close the business-IT gap even better, we provide a software component that stores all BO model representations, i.e. the business representation and all technical representations with detailed metadata in one repository—the BO Metadata Repository. It also stores all mappings and it is used by design time tools as well as runtime components of the Logistics Mall.

4.2.1 Objectives and Requirements

The objectives of the BO Metadata Repository are as follows:

- Support for Multilingual **BO Model Browsing** for Design Tools in the Logistics Mall, especially for the LPD (Logistics Process Designer [28]) and for App Development [33]
  - Search and browse structured metadata for the BO model, including internationalized names and descriptions of BOs and all their attributes, as well as of allowed enumeration values and their meaning
  - Display in simplified view (e.g., flattening of inheritance hierarchies)
- Further Support of **App Development**
  - Download of implementation files (XSD and Java archives) for the BO model
  - Management of mappings between the BO representations (domain model and technical representations as OAGIS-XSD, Java)
- Support of **Integration and Communication**
  - Check validity of BO instances
- Support of **BO Model Administration**
  - BO model versioning
  - Management/upload of structured metadata for a BO model version
  - Generation of (domain) documentation for the BO model version
  - Dependency analyses (which BO used which components and refers to which other BOs)

The BO Metadata Repository is able to store metadata of arbitrary UML models, since its stored metadata can be imported from any UML model of the UML tool used. This also eases the management of BO model metadata.

4.2.2 Architecture

The architecture of the Metadata Repository in principle follows the same architectural principles as described for the Instance Repository (cf. Fig. 23). This yields the same advantages for tool integration.

5 Background and Related Work

5.1 Business Objects

The provision of standardized business objects is not a new topic; therefore, we give an overview of existing approaches usable in the logistics domain.

It is highly important for a sophisticated BO model to have a good foundation by using an accepted standard, since Fraunhofer itself, although being neutral, is not a standards organization—thus we restrict our overview to existing standards. These standards have been developed over years by several domain experts and many experiences and corresponding feedback has been considered in these standards. This fact highly increases acceptance. The main conflict in the selection of a suitable standard is its widespread use in logistics versus its flexibility and thus the technical suitability for the Logistics Mall as part of a SOA and for flexible executed business processes.

There are three classes of e-business standards to consider:

- Process standards (to automate complex business processes) such as ebXML [34] and RosettaNet [35] with their implementation architectures are quite complex and mainly suitable for larger enterprises.
- Transaction standards (as the basis for the automation of business processes) like EDIFACT [20] (and EDIFACT subsets such as GS1 EANCOM), GS1 XML [22], UBL [36] and OAGIS [15], but also openTRANS [23] and the transaction part of RosettaNet appear to be suitable.
- Catalogue exchange formats (for electronic delivery of master data) may be needed additionally, if the process or transaction standard does not support this task.

Therefore, we focus on transaction standards. There are some comparisons of these standards from different views in [37–41].

UN/EDIFACT (United Nations Electronic Data Interchange For Administration, Commerce and Transport, ISO 9735) [20] initially developed in the 1980s is an important international multi-domain standard for logistics, since most existing logistics software provides some EDIFACT exchange interface. It is used by over 300.000 companies worldwide. However, it is technically outdated (delimiter-based) and thus not future-safe. It is inflexible, nearly not extensible, and targeted to fixed business processes. There are about 223 message formats (BODs) which are ASCII-encoded (not XML-based), and BODs are not distinguished from BOs. Because of its complexity, many industry-specific subsets of EDIFACT have been developed (EANCOM, EDIFICE, Odette, and others), and regularly two versions are produced each year. Due to its delimiter-based format, it is not well suited for a SOA.

The **UN/CEFACT** [14] e-Business stack consists of several modular specifications, namely CCTS and CCL: **CCTS** (Core Component Technical Specification [13], ISO 15000-5 and part 8 of ebXML) enables cross-organizational interoperability on data level. It is implementation-independent and includes a methodology for the development of context-specific semantic units and blocks of context-neutral templates for business vocabulary. An industry-specific adjustment of the data is performed by a context-driven principle for the generation of variants. Semantic syntax-agnostic building blocks represent general business data entities (context-free building blocks, common understanding by machine and human). The **CCL** (Core Component Library, TBG17) [42] is a repository for generic, CCTS-based business information. Core components from the CCL are also called ebXML Core Components and used in several eBusiness standards, e.g., UBL, RosettaNet, and OAGIS.

ebXML (Electronic Business using XML) [34] started 1999 is a family of XML-based standards sponsored by both OASIS and UN/CEFACT and adopted by ISO whose mission is to provide an open, XML-based infrastructure that enables the global use of electronic business information in an interoperable, secure, and consistent manner by all trading partners.

RosettaNet [35] is a non-profit consortium aimed at establishing standard processes for the sharing of business information (B2B), working to create and implement industry-wide, open e-business process standards. These standards form a common e-business language, aligning processes between supply chain partners on a global basis. RosettaNet is a subsidiary of GS1 US and has several local user groups. The European User Group is called EDIFICE. The RosettaNet standard is based on XML and defines message guidelines, interfaces for business processes, and implementation frameworks usable for interactions between companies. It focuses on the supply chain area, but also manufacturing, product and material data and service processes are in scope. The RosettaNet PIP directory consists of about 30 e-Commerce Message Types. In Europe, RosettaNet is not widely used, since EDIFACT is mainly used here.

openTRANS [23] from Fraunhofer IAO and the university of Essen is a small and XML-based transaction standard and can be seen as a complement to the catalogue interchange format BMEcat [43]. The scope of openTRANS is limited to eight transaction types: advice, invoice, order, order change, order confirmation, offer, receipt confirmation, request quote.

GS1 XML [22] from GS1 Germany exists since 2004, it is the successor of EANCOM, and the most used EDIFACT subset designed for Supply Chain Integration and therefore has the same focus. It is used by 20 industries in over 20 states

(2008), but mainly known in Germany—although not widely used here, the main stakeholders include Oracle, Siemens, Phillips, SAP, and DHL. It provides over 60 BOD types used in 12 sample scenarios and directly supports multi-step transactions, but does not include a separated BO model (BOs and BODs are not separated). The GS1 organization cooperation with UN/CEFACT, but—due to its nearness to EANCOM-only uses UN/CCTS methodology, not the UN/CCL components (e.g., [38] shows the difference in address formats). Instances are freely extensible (by so-called extensions), but the schema can only be extended by the GS1 organization following a formal process.

UBL (Universal Business Language) [36] from OASIS is a generic, open, and free, cross-domain, XML-based interchange format for business documents, which can be expanded to meet the needs of specific industries. UBL is based directly on UN/CEFACT CCTS, whose core components form the basis of 31 (in version 2.0) or 64 (version 2.1) business documents, thus it is less powerful than OAGIS especially for logistics. UBL was developed with the aim, among other things, to facilitate the exchange of data between companies belonging to different industries and therefore cannot use a common standard.

OAGIS (Open Applications Group Integration Specification) [15] from the OAGi is an international cross-domain transaction standard for B2B and A2A and exists since 1996 (only its first versions were not XML-based), it is used by over 38 industries in 89 states (05/2011); main stakeholders are IBM, Oracle, DHL, SAP, Microsoft. OAGIS 9.5.1 consists of 84 BOs, used in over 530 BODs (including master data exchange), the BODs are used in 64 sample scenarios, and OAGi provides Web service definitions. One of its explicit objectives is to provide a canonical business object model [27]. It integrates many other standards: UN/CEFACT, ISO, OASIS, CCTS/CCL and many more and can be used together with ebXML and is EDIFACT-compatible. The OAGi quickly adopts modern trends, e.g. soon the JSON exchange format will be supported additionally to XML to better support mobile devices, and there are cloud and BPMN initiatives. Most important for us is its openness and schema extensibility by XSD overlays, in addition to instance extensions by freely usable so-called user areas.

GDTs (Global Data Types) [44, 45] from the SAP AG is a proprietary and not freely usable business object model also based on CCTS. SAP has successfully shown that a global business object model stored in its Enterprise Service Repository is very beneficial for tool and data integration: the model is used by many of its applications and components.

5.2 Bridging the Business-IT Gap

Bridging the gap between business and IT is not a new topic: one approach is composite applications [9], each consisting of a stable service layer and a rather flexible business process layer, which defines the application. They connect the IT and business layer for each application.

Another new trend is smart process applications [46] and smart process apps platforms, a new generation of BPMS (business process management suites). Most big BPM providers have it on their agenda or provide first products, evaluated by Forrester in [47]. The borders between apps and BPM system become fuzzy, since the apps more and more use functionality of the BPM platform. These apps are highly flexible, loosely structured, easily usable and mainly intended for human interaction by business people ("Designed for People, Build for Change, Critical to Success").

In [48] some important underlying techniques, namely front-loading and look-ahead are introduces to minimize the business-IT gap for BOs (data), and for services and processes (behavior), therefore we summarize it here:

Front-Loading means that the domain user provides additional information that he knows and that would otherwise be missing for the implementation. We instead use pre-implementations that may be provided by IT experts of Logistics Mall service providers, not implementations derived automatically from front-loading information. Front-Loading includes aspects like organizational units and roles, BOs, domain-specific services information, flexibility points (where e.g. an exchange of services is expected), foreseeable exceptions, form specifications (to derive GUIs), etc.

Look-Ahead means to find and reuse building blocks of all kinds (which already have those Front-Loading aspects), i.e. pre-implemented building blocks for data, behavior, exceptions and more. That is our focus, since the Logistics Mall business model is a shop of pre-implemented building blocks, combinable with correctness-by-construction in mind.

BOs (Data): Executable process models need specifications of all data objects (or BOs) to precisely define the interface of the process model and its internal data flow, i.e. for starting process instances, calling services and for user interaction via GUIs. In process modelling normally only the name and the most relevant attributes of data types are specified. Using standardized BOs, all additional implementation information is already available.

- **Front-Loading**: Only domain experts have the knowledge of required domain information and its structure, formats and codes. Using standard BOs, the data analysis has already been done by several domain experts.
- **Look-Ahead**: Existing data object types should be reused when designing business process models. Using the BO model, there already exist detailed specifications (documented UML class diagrams and structured data from the BO Metadata Repository) and even implementations (Java classes and the BO Instance Repository). To enable reuse of existing BO types through look-ahead, these are easily accessible to business process designers via the LPD tool [28]. As an advantage, efforts for designing, specifying, and implementing BOs are significantly reduced [12]. Even more important, business process models then refer to the same data types (BOs) as existing services and human tasks. Due to this harmonization, the overall efforts for implementing the transformation of a data structure defined at business level into another one used at technical level can be avoided [12] during service calls and process execution.

Services (Behavior): Several activities in a process model use services (or apps/applications) for their implementation. During runtime, process instances have to communicate with these services via exchange of data objects and events, the latter in asynchronous situations. Thanks to BPMN 2.0, there is no more need to bridging the gap between business process models and system process models or service composition specifications like BPEL, as described in [49], this narrows the business-IT gap already significantly.

- **Front-Loading**: In general, for each service to be used in a process model, its functionality and the BOs of its input and output parameters need to be prespecified, and which IT system shall offer this service. Using the LPD, a business process modeler may then use this information. Different systems may store different attributes for BOs, i.e. those they need. To foster front-loading, the Logistics Mall MMP [33] supports describing offered apps in business view that is pre-linked to the technical descriptions needed for service execution, and service governance is defined. Technical descriptions like WSDLs and XSD data type specifications are used, but they are not comprehensible to business process designers having no or only limited IT skills and are therefore hidden from them.
- **Look-Ahead**: Usually, service descriptions and service operations are published in a repository (SOA repository or LDAP in our case). To enable look-ahead, a service should be described in a language (or graphical notation) easily understandable to business process designers [48], as described above. Furthermore, it must be easy to search for needed services and to access and understand related descriptions. Based on this, business process designers can find appropriate services and use them in corresponding process steps. Existing technical service specifications are pre-linked with these business level specifications to avoid unnecessary implementation steps. Reuse of existing services reduces efforts and costs for service implementation or service renting. As a disadvantage, adjustments of the defined process logic to the available service set might become necessary.

6 Conclusion

We presented the new Logistics Business Objects model BO4L developed for and used by the Logistics Mall–its objectives, design, development methods, supporting software and how it fills the business-IT gap.

BO4L is a Logistics domain model developed by a team of Logistics experts as an abstraction of a canonical, flexible, modern, and lively technical standard, namely OAGIS. By this construction, the domain model is highly standards-based and independent of specific applications and good acceptance in the Logistics domain is predictable, becoming the "new oil" of logistics.

By supporting a simple business view and a tightly coupled complex standards-based technical view, this standard reduces horizontal gaps in two dimensions: it is a common language for Logistics experts of different cooperating parties and a common technical transaction standard for IT systems of all partners, and it is used for tool integration of Logistics Mall offers.

Additionally, by its construction as an elaborated abstraction of a technical standard, the Logistics BO model is also capable to narrow the vertical business-IT gap, both on data level and on activity level. The latter is used for a new paradigm in business process modelling.

The Logistics BO model is not only a data model with extensive documentation, but also consists of many technical artefacts, i.e. different technical representations of all business objects and also of software components supporting the handling of business objects at design time and at runtime.

We only sketched the main ideas of app integration, standard-based app communication, and app orchestration by process models, all with a focus on business objects. For a detailed discussion on how applications or apps get Mall-conformant BO-based interfaces or how process models are modelled and executed in the Logistics Mall we refer to [33] and [28] respectively.

References

1. Fraunhofer Gesellschaft: Business objects for logistics. http://www.bo4l.org (2013)
2. EffizienzCluster LogistikRuhr. http://www.effizienzcluster.de/en
3. Fraunhofer Gesellschaft: Logistics Mall. http://www.ccl.fraunhofer.de/en.html (2013)
4. Holtkamp, B., Steinbuss, S., Gsell, H., Loeffeler, T., Springer, U.: Towards a logistics cloud. In: 2010 Sixth International Conference on Semantics, Knowledge and Grids. http://ieeexplore.ieee.org/stamp/stamp.jsp?arnumber=05663532 (2010)
5. Bundesministerium für Wirtschaft und Technologie: Exponat „IKT-Innovationen in der Logistik – Wachstumsmotor für Unternehmen", Nationaler IT Gipfel. http://www.post-und-telekommunikation.de/PuT/1Fundus/Dokumente/7._Nationaler_IT-Gipfel_2012_Essen/20_it-gipfel-2012-exponat-ikt-innovationen-in-der-logistik,property=pdf,bereich=itgipfel,sprache=de,rwb=true.pdf (2012) (in German)
6. Nationaler IT Gipfel. http://www.it-gipfel.de
7. Hinssen, P. (ed.): Information is the new oil—drilling new sources of innovation, across technology. http://datascienceseries.com/assets/blog/GREENPLUM_Information_is_the_new_oil-LR.pdf
8. Nettsträter, A., Geißen, T., Witthaut, M., Ebel, D., Schoneboom, J.: Logistics Software Systems and Functions: An Overview of ERP, WMS, TMS and SCM Systems. Springer, Heidelberg (2013)
9. Stiehl, V.: Composite application systems—systematisches konstruieren von verbundanwendungen unter verwendung von BPMN, dissertation, TU darmstadt. http://tuprints.ulb.tu-darmstadt.de/2751/ (2011) (in German)
10. DIN German Institute for Standardization e.V.: Economic benefits of standardization—Summary of results—Final report and practical examples, Beuth Verlag. http://www.din.de/sixcms_upload/media/2896/economic_benefits_standardization.pdf (2000)
11. Berlecon research: E-business standards in Germany—current state, problems, prospects. http://www.berlecon.de/studien/downloads/Berlecon_eStandards2010_ExecSum.pdf (2010)
12. Böhmer, M., Daniluk, D., Schmidt, M., Gsell, H.: Business object model for realization of individual business processes in the logistics domain. In: Clausen, U., ten Hompel, M., Klumpp, M. (eds.) Efficiency and Logistics, pp. 237–244. Springer, Berlin (2013)

13. UN/CEFACT: Core components technical specification, version 3.0. http://www.unece.org/fileadmin/DAM/cefact/codesfortrade/CCTS/CCTS-Version3.pdf (2009)
14. UN/CEFACT. http://www.unece.org/cefact.html
15. Open Applications Group, Inc.: OAGIS release 9.5.1. www.oagi.org (2012)
16. Schmidt, M., Tilp, M., Böhmer, M.: Fachliche dokumentation der geschäftsobjekte für die logistik, release 1.2.2 "blaze", technical report, fraunhofer gesellschaft (2013)
17. Scott-Sabic, V.: Dictionary of Logistics. Deutsch-Englisch/Englisch-Deutsch. Springer, Heidelberg (2004)
18. Heidenblut, V., ten Hompel, M.: Taschenlexikon Logistik. Springer, Heidelberg (2011). (in German)
19. GS1: GS1 General Specifications. http://www.gs1.org/docs/gsmp/barcodes/GS1_General_Specifications.pdf (2014)
20. UN/CEFACT: Introducing UN/EDIFACT. http://www.unece.org/trade/untdid/welcome.html
21. Open Applications Group, Inc.: Business object document (BOD) message architecture for OAGIS© release 9. + , OAGi white paper, version 1.0. http://www.oagi.org/oagi/downloads/ResourceDownloads/2011_0408_BOD_Message_Architecture_V9x.pdf (2011)
22. GS1: GS1 XML. http://www.gs1.org/ecom/xml
23. Fraunhofer IAO: openTRANS. http://www.opentrans.de
24. Open Applications Group, Inc.: User's guide for extending OAGIS 8, OAGi reference guide, version 1.0. http://www.oagi.org/oagi/downloads/ResourceDownloads/Extending_OAGIS_8.pdf (2003)
25. Open Applications Group, Inc.: Extending OAGIS 9.x through overlays—OAGi cheat sheet. http://www.oagi.org/oagi/downloads/ResourceDownloads/Extending_OAGIS_v9.doc (2009)
26. Open Applications Group, Inc.: OAGi development methodology, OAGi white paper. http://www.oagi.org/oagi/downloads/developmentmethodology/2013_OAGi_Development Methodology.pdf (2013)
27. Open Applications Group, Inc.: OAGIS—A "Canonical" business language providing both vertical and horizontal requirements, OAGI white paper, version 1.0. http://xml.coverpages.org/OAGIS-Canonical.pdf (2002)
28. Bochon, I., Ivens, V., Nagel, R.: Challenges of Cloud Business Process Management. Springer, Heidelberg (2013)
29. Object Management Group, Inc.: Business process model and notation (BPMN), version 2.0. http://www.omg.org/spec/BPMN/2.0 (2011)
30. Steinbuß, S., Weißenberg, N.: Service design and process design for the logistics mall cloud. In: Yang, X., Lu, L. (eds.): Principles, Methodologies, and Service-Oriented Approaches for Cloud Computing. pp. 185–205, IGI Global, Hershey (2013)
31. Carr, N.G.: IT doesn't matter. Harvard Bus. Rev. http://hbr.org/2003/05/it-doesnt-matter/ar/1 (2003)
32. Date, C.J.: An Introduction to Database Systems. Addison-Wesley Publishing Company, Englewood Cliffs (1977)
33. Eggemann, J., Leveling, J., Weiß, N.: Business Apps Meet the Challenge of Covering Continually Changing Logistics Requirements. Springer, Heidelberg (2013)
34. OASIS: ebXML. http://www.ebxml.org
35. RosettaNet. http://www.rosettanet.org
36. OASIS: Universal business language TC. https://www.oasis-open.org/committees/ubl/faq.php
37. Liegl, P., Zapletal, M., Pichler, C., Strommer, M.: State-of-the-art in business document standards, industrial informatics, pp.234–241. http://publik.tuwien.ac.at/files/PubDat_186928.pdf (2010)
38. Kabak, Y., Dogac, A.: A survey and analysis of electronic business document standards. ACM Comput. Surv. **42**(3), 1–31 (2010). https://lists.oasis-open.org/archives/set/200909/pdf00000.pdf (2010)
39. Melleri, I., Hiekkanen, K. ,Mykkanen, J.: comparing information models in XML-based e-business standards—a quantitative analysis. In: IEEE 15th International Enterprise Distributed Object Computing Conference, pp.225–234 (2011)

40. Lampathaki, F., Mouzakitis, S., Gionis, G., Charalabidis, Y., Askounis, D.: Business to business interoperability: a current review of XML data integration standards. Comput. Stan. Interfaces **31**, 1045–1055 (2009)
41. Li, H.: XML and industrial standards for electronic commerce. Knowl. Inf. Syst. **2**(4), 487 −497 (2000)
42. UN/CEFACT: Core components library. http://www.unece.org/cefact/codesfortrade/unccl/ccl_index.html
43. BMEcat. http://www.bmecat.org/English/index.asp
44. SAP: SAP Netweaver CE 7.2 documentation, global data types. http://help.sap.com/saphelp_nwce72/helpdata/en/b8/870116599c48a4aa2bb15b7a9967da/frameset.htm
45. SAP: Global data types. http://wiki.sdn.sap.com/wiki/display/GDT/SAP+Global+Data+Types+(GDTs)
46. Bartels, A., Moore, C.: Smart process applications fill a big business gap, forrester report (2012)
47. Bartels, A., Moore, C.: Smart Process Applications, Forrester Wave (2013)
48. Bauer, T., Buchwald, S., Reichert, M.: Improving the quality and cost-effectiveness of process-oriented, service-driven applications: techniques for enriching business process models. In: Service-Driven Approaches to Architecture and Enterprise Integration, pp.104 −134. http://dbis.eprints.uni-ulm.de/936/ (2013)
49. Buchwald, S., Bauer, T., Reichert, M.: Bridging the gap between business process models and service composition specifications. In: Lee, J., et al. (eds.) Service Life Cycle Tools and Technologies: Methods, Trends, and Advances, pp. 124−153. IGI Global, Hershey. http://dbis.eprints.uni-ulm.de/713/1/BookChapter_BRB2011.pdf (2012)

Challenges of Cloud Business Process Management

Iryna Bochon, Volker Ivens and Ralf Nagel

Abstract Cloud Business Process Management means more than outsourcing of the Process Editor and the Execution Engine. The overall Process lifecycle has to be adopted to face the challenges of Cloud Computing. In this chapter we discuss these challenges and present our solutions to reconcile cost efficiency, security, availability, agility and low administration overhead.

Keywords Business process management · Cloud computing · Logistics mall · Logistics process designer

1 Introduction

In today's world you need not buy your own vehicle if you want to travel by car. An alternative would be to join a car sharing community. In car sharing you are not responsible for buying and servicing your own car, you just rent a car from the car sharing pool for the time you need it. The decision which solution suites you best is based on different influence factors like the price list, mileage, travel behavior or the distance to the next pick up point.

The main aspect of Cloud Computing is in analogy to car sharing the change of view from *owning* an IT System to just *renting* and *using* it. Instead of giving a definition of the term Cloud Computing we want to examine some of the key factors behind this approach.

I. Bochon (✉) · V. Ivens · R. Nagel
Fraunhofer Institute for Software and Systems Engineering ISST,
Emil-Figge-Str. 91, 44227 Dortmund, Germany
e-mail: iryna.bochon@isst.fraunhofer.de

V. Ivens
e-mail: volker.ivens@isst.fraunhofer.de

R. Nagel
e-mail: ralf.nagel@isst.fraunhofer.de

M. ten Hompel et al. (eds.), *Cloud Computing for Logistics*,
Lecture Notes in Logistics, DOI 10.1007/978-3-319-13404-8_7

- *Elasticity*: The possibility to short-dated add or remove computing resources (CPU, memory, storage capacity) gives the Cloud customer a flexibility which hasn't been possible in the past by obtaining own hardware or the classical outsourcing of IT services. This advantage for the customer is tightly connected with the problem of a precise forecast of the computing resource demands by the cloud operator. Different approaches may lead to overbooking problems at peak times in a Cloud data center.
- *Security*: The security problem populated by the mass media nowadays seems to be overrated and applies more to free of charge Cloud services like Dropbox[1] or most Google services.[2] In the case of business appliances in Europe data protection is severely regulated by German and European law. The implementation and monitoring of these rules is essential for every serious Cloud operator.
- *Cost reduction*: This is another factor often presented by popular publications. In reality it is quite hard to compare the cost of classic IT management with Cloud computing. What is the equivalent value of Cloud elasticity and flexibility? Neglecting these facts is comparing apples and oranges. For the Cloud operator it is easier to balance the utilization of the hardware between multiple customers and to absorb performance peaks of a single customer. If peaks of all Cloud customers appear simultaneously a real cost reduction may be unachievable.
- *Remote Access*: Cloud Computing resources (IaaS,[3] PaaS,[4] SaaS[5] [1]) are accessed through a network (not necessarily the public Internet). That means the physical computing center behind the Cloud may be located anywhere around the world. This potential advantage must be balanced to the fact, that the legal restrictions over the world are quite heterogeneous. Data protection and governmental control differ in various parts of the world [2].
- *Availability*: In most businesses a high availability of Cloud Computing resources is required. Any single point of failure could damage the overall availability extremely. All involved components in Cloud Computing like servers, storage, networking, operating systems, power supplies, cooling and so on may lead to fatal blackouts and thus must be kept failsafe within the desired specifications.

Looking at the advantages of Cloud Computing it is obvious to think about a combination of Cloud Computing and Business Process Management (BPM). Business Processes are ubiquitous in our enterprise world. Sometimes they are hidden in standard operating procedures. A better way is to use any kind of

---

1 http://www.dropbox.com.

2 Google Mail, Drive, Picasa and so on.

3 IaaS: Infrastructure as a Service.

4 PaaS: Platform as a Service.

5 SaaS: Software as a Service.

Business Process Notation like EPC,[6] BPEL[7] or BPMN.[8] Using a Business Process Notation for the execution and monitoring of processes is the state-of-the-art technology in driving an enterprise. With BPMN 2.0 there exists a standardized notation including an executable semantic. Various tools and execution engines support this standard released in 2011 by the OMG [3].

BPMN is an evolution of other basic modeling concepts and languages like Petri nets or state machines. There are scientific researches proving that all BPMN graphs can also be modeled as Petri nets [4]. Those graphs do not contain execution information, so in order to execute a Process model there has to be a special description language for the execution. One language established for that purpose was the Business Process Execution Language, which describes the orchestration of Web services [5] included in a Process model. One of the weaknesses of this language was the missing support for human activities, so some extensions of BPEL including human tasks emerged (BPEL4People). For some time the combination of BPMN for modeling and BPEL variants for execution was used until BPMN 2.0 was finally released.

Typically domain specialists design Process models, but in most cases they do not have the needed technical experience and understanding of the infrastructure or services for the execution of a process. Consequently, a designed Process model needs some modifications and technical refinements performed by IT specialists to make it executable. For the execution of a Process model every activity must be mapped to a service implementation (e.g. a Web service) in order to provide the desired functionality. Additionally, the outputs of activities have to be wired with the inputs of successive activities. Incoming external event have to be delivered to corresponding Process instances. Outgoing events must be delivered to the correct receiver.

To execute a Business Process model, there have to be several technical refinements to the composed model. Process internal and external events must be wired correctly for the execution of the model.

An executable Business Process Notation covers the following key aspects:

- *Control flow*: the order in which the Process model elements are executed (Activities, Gateways, …).
- *Data structure*: the definition of data stored in the Process context and exchanged with services.
- *Data flow*: describes the flow of data between Processes and services.
- *Messages and Events*: specialization of the data structure and flow for communication and synchronization between Processes and services.

---

[6] EPC: Event-driven process chain.

[7] BPEL: WS-Business Process Execution Language.

[8] BPMN: Business Process Model and Notation.

- *User and Roles*: specialization of the data structure representing the organization chart of peoples interacting with the Process.
- *Application landscape*: the applications providing the service implementations for the activities.

A typical way to exchange data between Process activities is the use of a Process context. A Process context works like shared memory for all activities in the Business Process. Data objects can be stored in Process variables and every activity can access or update it. Required input and output objects of a Process activity must be available in the Process execution context in order to execute this Process.

Some typical problems with shared memory concepts are:

- *Data compatibility*: the data exchanged between services must be compatible for successful execution.
- *Context size*: a growing size of Process context may lead to performance and scalability problems.
- *Transactions*: in concurrent Processes multiple instances or process threads may try to alter the same data at the same time. This may lead to inconsistency in the data and transaction methods are needed to control the modifications.
- *Synchronization*: the synchronization of data duplicated from Process context to external data sources may also lead to data inconsistencies.

2 Cloud Business Process Management (CBPM)

Cloud Business Process Management means more than moving of a Process Editor and an Execution Engine into a Cloud infrastructure. Although it is possible to use IaaS or PaaS to operate a BPM infrastructure in the Cloud, virtualization of these servers does not give a big advantage over classic IT management. Hence, in the following we focus on SaaS based solutions.

CBPM should focus on *creating (modeling)* and *using* Processes rather than *implementing* and *operating* the services. Therefore, we see a clear separation of duties between the two stakeholders involved:

- The *Cloud operator* is responsible for running and maintaining the Business Process environment in the Cloud together with the connected SOA[9] services. This includes setting up new or updating existing services and applications.
- The *(Cloud) customer* of CBPM designs, tests and uses their Business Processes orchestrating the selected Cloud services.

As a consequence from this separation of duties, privileged accounts for the underlying operating systems, databases and so on are not issued to the Cloud customer. Even the execution of foreign or uncertified code by the Cloud operator

[9] SOA: Service Oriented Architecture [6].

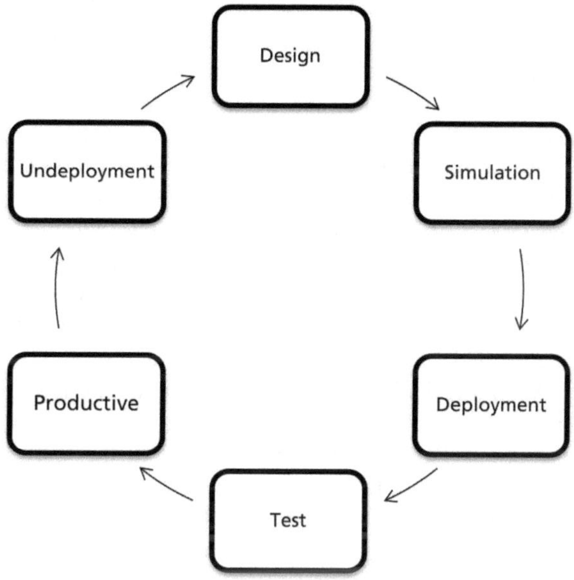

Fig. 1 Business Process Lifecycle

may be legally problematic and dangerous. Uncertified code provided by the customer could harm the system health of the Cloud infrastructure or contain vulnerabilities of any kind. Thus a Business Process designed and used in the Cloud may only consist of predefined and by the Cloud operator certified building blocks.

Designing a Business Process goes through some development steps, well known from software development techniques (cf. Fig. 1). In the beginning a Process is typically *designed* in a graphical notation. This step is iteratively performed and can be accompanied by internal *simulation* phases, which help the designer evaluating his newly created or altered Process model. The design phase is usually followed by a *test* phase. During the test phase a Process model can be executed. If necessary the Process model and depending artifacts first have to be *deployed*. In most Execution environments the deployment of new Process models is simply based on XML files. More complex is the preparation of all services referenced by a Process model. If a service hasn't been used by a Process model before, it must be installed and configured before usage. Often the Process model has to be updated to reflect the current IP or URL of the newly deployed service.

An execution of a Process in the test phase must be clearly marked preventing misinterpretations of Process data and incoming or outgoing signals to external systems. Any error found while testing may lead to another design phase. After a successful test phase the Process changes to the *productive* phase and can be used as desired. Any errors found in the productive phase will also lead to a new design phase. At the end of its lifetime a Process model is *undeployed*, saving computing resources and preventing the creation of new Process instances.[10] The physical

[10] Instance of the Process model.

undeployment of a Process model usually has to wait until all running Process instances have been completed. Through the agility of this method a newer version of a Process model may be deployed, tested or live executed, even while the predecessor version is still in operation. Therefore Business Process Management needs a built in Version management for Process models.

Additional support is required from a Process's services and applications in different phases of the Business Process Lifecycle. Typically for testing purposes special test instances of applications are used to distinguish between productive and test data. Demanding a cost reduction in the Cloud, this duplication of systems is undesirable. A better way is to find a way to mark Process instances and corresponding data as test data using only one service instance during test and productive phases. This allows also merging test and productive execution at the same time.

With the lack of privileged accounts for the Cloud customer the Execution environment and Process services have to support the monitoring of a Process execution. This monitoring is needed in the test phase to make sure everything works as desired. During the productive phase the monitoring is required to analyze any problems which might occur during Process execution. While a broken Process instance may be ignored during a test phase, in productive mode the Cloud customer has to deal with these instances. For the handling of broken Process instances a specialized repair function must be provided by the Cloud environment. With the help of these functions corrupted data must be updated and a Process execution is reset into a prepared state so that the execution can be continued.

Another challenge in CBPM is to connect the Cloud environment with external systems. Only considering Cloud based systems will narrow the application area. Especially in the Logistics domain a lot of physical hardware outside the Cloud has to be included into a Business Process. Starting from barcode scanning and printing, a Cloud process must communicate with external Enterprise Resource Planning tools or Warehouse Management Systems. A specialized Gateway has to deal with incoming and outgoing messages and data, assigning incoming data to the corresponding Cloud customer and Process instance.

We can summarize the key challenges of CBPM as follows:

- *Elasticity and SaaS*: this outlines the Cloud Computing aspect of CBPM ("just use" metaphor extended to processes and connected services).
- *Executability of building blocks*: CBPM assembles only executable and certified building blocks. These building blocks especially support the different Business Process Lifecycle phases with monitoring and repair functionalities.
- *Compatibility*: the compatibility of data exchanged by building blocks must be guaranteed.
- *Connectivity*: with the help of a Cloud gateway [7] the communication to the outside world can be established. The Gateway configuration is a part of the Process definition.

3 Logistics Process Designer (LPD)

3.1 Architecture of the LPD

The Logistics Mall is a marketplace for Logistic services and applications [8]. Vendors of Logistic services and applications can offer their products as monolithic applications or specialized services. This allows users of Logistic or IT services to rent only required software parts for their enterprise. It is also possible to create individual Logistic Process models, built from available services, and execute them in the Cloud.

The development of the Logistics Mall was divided into three stages (cf. Fig. 2). In the first stage only monolithic applications were supported. Customers could work with their booked applications (like Software as a Service) inside a tenant specific Customer Access Framework (CAF). For the second stage the Logistics Mall Cloud infrastructure was extended with an Enterprise Service Bus[11] (ESB). The ESB allows interoperability between the booked applications in the Cloud. To harmonize the communication between applications of different vendors, a set of standardized Business Objects (BOs) [10] has been developed, together with a Repository able to store BOs. The last stage of the Logistics Mall aims to provide support for *Business Process as a Business Service* (BPaaS).[12] Monolithic applications from stage two are replaced by small services that can be arranged to Process models. The specialized tool for building Business process models in the Logistics Mall is the Logistic Process Designer (LPD).

The Logistics Process Designer is a Web based Process modeling tool which uses an independent Business Process Notation (cf. Fig. 8). Target user of the LPD is every domain specialist without the need for deep IT knowledge. The graphical notation used is intuitive for beginners and easier to learn than most subsets of BPMN. The LPD uses only pre-implemented building blocks making a subsequent refinement unnecessary. Building blocks of the LPD can be translated into a selection of Business Process Languages like BPEL or BPMN 2.0, making it possible to use available Open Source or commercial Business Process Execution Engines.

As shown in Fig. 3, we separate three different levels: the top level consists of the *Process Modeling Taxonomy Editor* (*PMT Editor*), the *Logistics Process Designer* (*LPD*) and the *Process Repository Frontend*. With the PMT Editor the operator of the Logistics Mall manages the Process Modeling Taxonomy containing all available building blocks. With the LPD the building blocks from the taxonomy[13] are composed into Process models. Finally, the Business Process Lifecycle

---

[11] "An Enterprise Service Bus is a standard-based integration platform that combines messaging, web services, data transformation and intelligent routing in a highly distributed, event-driven Service Oriented Architecture" [9].

[12] In Analogy to Software as a Service.

[13] A taxonomy is a systematic and hierarchical classification of things [11].

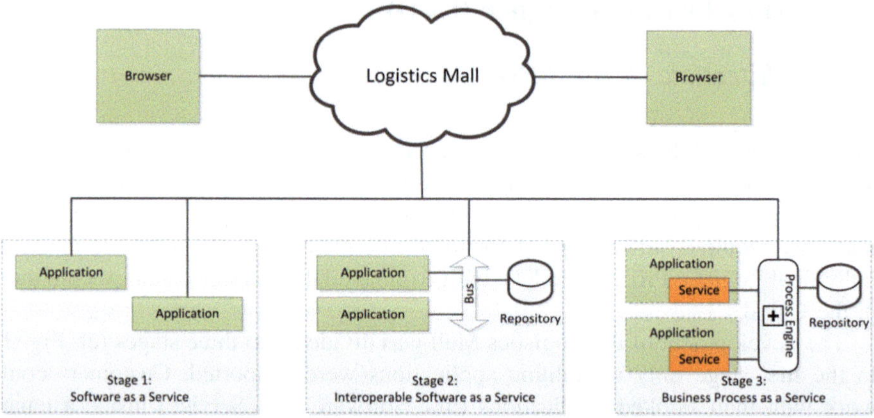

Fig. 2 Logistics Mall stages 1–3

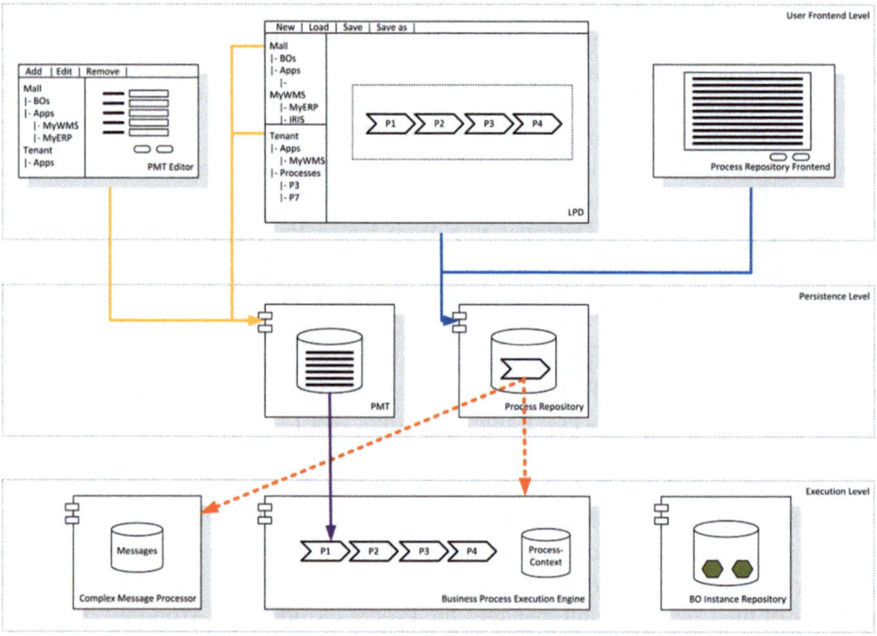

Fig. 3 Big picture: Process Management in the Logistics Mall

of these Processes is controlled with the Process Repository Frontend. The access to all mentioned Web applications is granted through the integrated Logistics Mall SSO Gateway.

The middle tier or *Persistence level* contains the two essential data sources: The *Process Modeling Taxonomy* stored in a LDAP[14] server and the *Process Repository* which is an ordinary RDBMS.[15] The Process Repository stores various Process models of multiple tenants. Each model can be stored in different versions identified by a simple version number. A special version is the *head version* which represents the latest version of this model, saved by a user. In difference to the head version, all other versions of a model are immutable. Only these versions can be tested or deployed. A head version has to be tagged (with a version number), before the Business Process Lifecycle controlled by the Process Repository Frontend is started.

On the *Execution level* we find all systems required for the execution of our Processes. Key part in the execution is the *Business Process Execution Engine*. For the Logistics Mall we use the Activiti Open Source BPMN 2.0 engine [12]. A Process model stored in the Process Repository is translated into an XML file according to the BPMN 2.0 specifications. The code generator is controlled by the Process Repository Frontend and customized for the Activiti engine. The translation uses three main input templates:

- The *Base template* is built into the code generator and reflects any specifics of the target Execution Engine platform.
- Every building block in the Logistics Modeling Taxonomy has a custom *Execution Block template*. For a Process the block templates of all used building blocks are joined with the base template.
- The values of *Building Block parameters* defined in the Process model are extracted from the process description in the Process Repository and are merged with the rest of the template.

The integrated Business Process Execution Engine can easily be replaced by competitive engines. The code generator is prepared for customization simply because the three template levels are exchangeable. Beside BPMN any executable notation and suitable execution engine is feasible (e.g. a BPEL engine).

The two other systems in the Execution level are the *Business Object Instance Repository* [10] and a *Complex Event Processor* (*CEP*). The BO Instance Repository is a specialized persistence mechanism for the Business Object Model of the Logistics Mall [10]. The CEP is the event gateway for running Process models. With the help of the CEP it is possible to interoperate with the Process engine. Basic operations of the CEP are:

- Starting a new Process instance.
- Notify a waiting Process.
- Stop a running Process (Cancelation).

[14] LDAP: Lightweight Directory Access Protocol.

[15] RDBMS: Relational Database Management System.

Additionally it is possible to extract data from a message event. With this data it is possible to:

- Create new Business Objects in the repository.
- Update existing Business Objects.
- Mark a Business Object as obsolete.

Finally events to the CEP can be sent from various origins. The following list gives a brief overview of systems able to send events to the CEP:

- Running Process instances.
- The BO Instance Repository.
- External systems (through the Logistics Mall gateway).
- User interfaces.
- Services or Logistics Mall Apps [13].

All events handled by a Process model have a description in the tenant-specific part of the PMT. Therefore in the LPD events are treated like any other activity from the taxonomy when building a Process model.

The Logistics Process Designer is a GWT [14] application using HTML 5 [15] elements, running on an *Apache Tomcat*[16] Web server. HTML 5 is still in development but it is the future standard for modern Web pages, featuring many new functions for Browsers without the need for Plug-ins. Its architecture is shown in Fig. 4. The LPD Frontend is the client side of the LPD which displays the user interface. The Canvas of the LPD is based on the *yFiles for HTML* Framework [16] to render graphs and the GWT-DND[17] library for drag and drop features. The frontend communicates with the Persistence and Process Modeling Taxonomy (PMT) backend services. The LPD Persistence uses *JPA*[18] to access a *PostgreSQL*[19] Database. The Persistence API provides the functionality to save and load Process models as well as versioning them. Among other data, the Process Modeling Taxonomy stores meta information about the tenant-specific Process models and administers the IDs by which the models are stored in the Persistence module. The PMT is stored in an *OpenLDAP*[20] server and is accessed through *JNDI*.[21] The PMT includes all available Logistics Mall applications and services combining it with the information's about the applications and services of a tenant. *Spring*[22] is used for configuration and integration of the components Frontend, Persistence and PMT. The Common component contains the data transfer objects for the communication of the core components.

---

[16] Apache Tomcat: http://tomcat.apache.org/.

[17] GWT-DND: Drag & Drop Library for GWT: https://code.google.com/p/gwt-dnd/.

[18] JPA: Java Persistence API.

[19] PostgreSQL: http://www.postgresql.org/.

[20] OpenLDAP: http://www.openldap.org/.

[21] JNDI: Java Naming and Directory Interface.

[22] Spring Framework: http://www.springsource.org/.

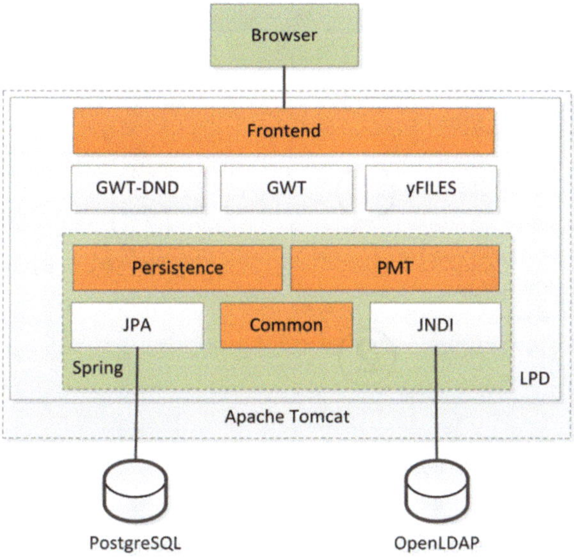

Fig. 4 LPD modules and architecture

The LPD interface (Fig. 5) can be separated into three areas, the tenant independent taxonomy tree (1 + 2), the tenant specific tree (3 + 4) and the canvas (5). The canvas is the modeling area, where the Process models are designed. The tenant independent taxonomy tree contains the Control flow elements as well as all Mall applications and services. The Control flow elements are shown as number 1. These elements are the building blocks to describe the control flow or events. The other entries of this tree are Logistics Mall Applications, such as the Nelsan Shop [17] in Version 1.1.0 with the Service "Print delivery note" (2).

The tenant specific taxonomy tree lists all applications the tenant has already booked, as well as those he is planning to book in the future. The state of an application (3) is indicated by its coloring, this gives an overview which applications a tenant needs to acquire to run all of his processes. Number 4 shows a list of Process models the tenant has already created. They can be used as Subprocess building blocks in other models.

The PMT is the key part of the LPD. Every building block used in a Process model is taken from here. The PMT provides several functions such as classification, hierarchical structure and management of the available building blocks. Additional meta information is stored for each building block. The user picks a building block from the taxonomy and drags it onto the canvas. Afterwards the blocks can be customized and connected with edges.

In the LPD the PMT is represented by two tree widgets.[23] The PMT structure consists of multiple root categories, which are connected by references in between. One part of this taxonomy has to be administered by a *Taxonomy administrator*, the

[23] A widget is a graphical component of the GUI, like a textbox or tree.

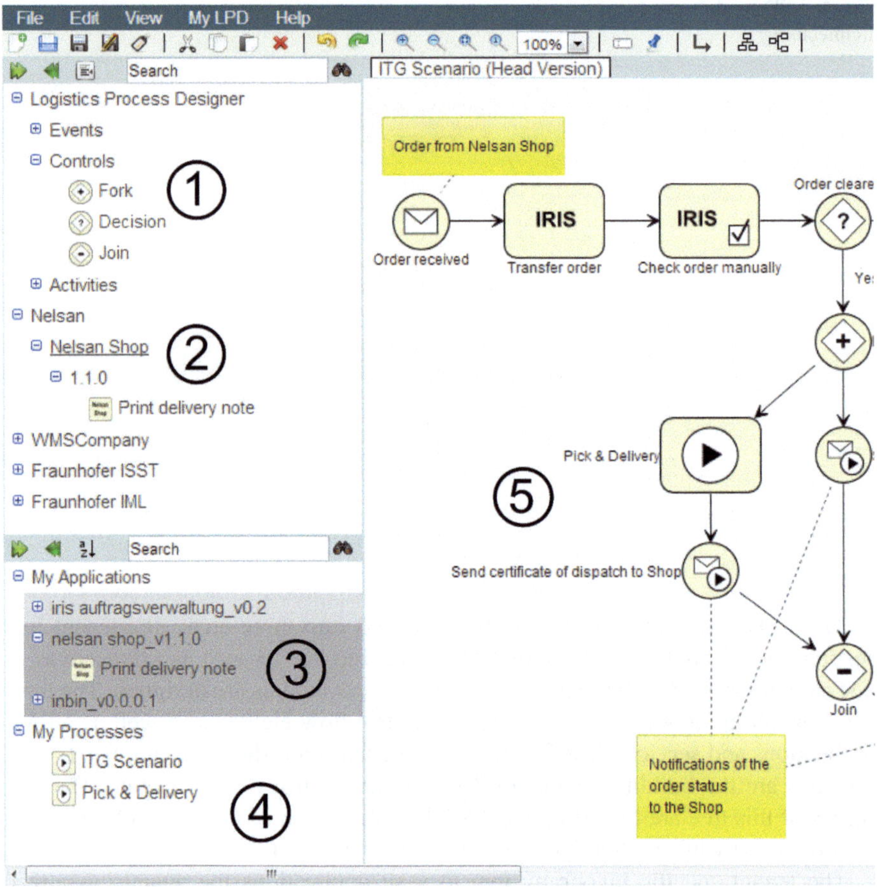

Fig. 5 LPD interface components

other part is changed programmatically during the design of a Process model. The PMT has a tree structure which allows a deep hierarchy for the objects. Two groups can be identified in the taxonomy, one part is *tenant independent*, the other is *tenant specific* (Fig. 6).

The independent part consists of trees for *Logistics Mall Business Objects* [10], *Control flow elements*, *Logistics Mall applications* and *configurations*. The application tree contains information about the provider of an application, which versions exist and which services this application version offers. The Business Objects required as inputs or outputs for these services are described as well. Execution templates for multiple target platforms are stored for each activity of an application. An Execution template is the link between the building block in the canvas and the execution code in the Execution engine. The code generator uses Execution templates when translating a Business Process into an executable Process notation. Template parameters are used to replace special placeholders within Execution

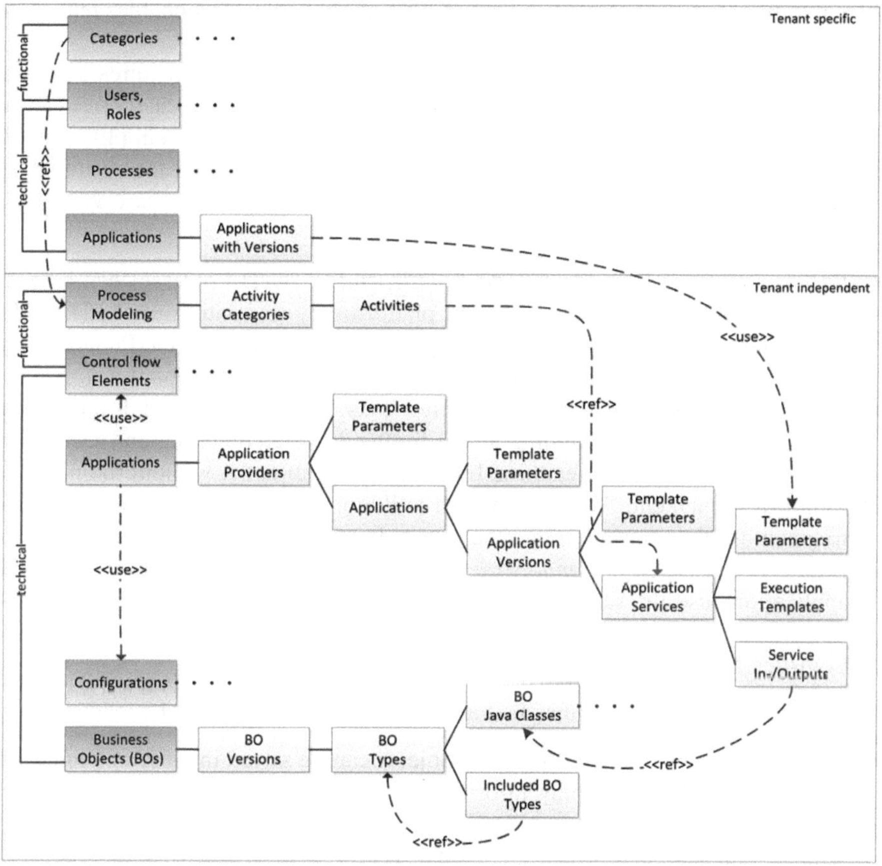

Fig. 6 Process model taxonomy

templates. They can be defined on any level in the application tree. Through inheritance these parameter get passed down to the application services level. The *Logistics Mall Business Objects* tree represents all predefined Business Objects in different versions from the Logistics Mall Business Object model which are referenced as application service inputs or outputs. The Control flow elements tree contains several control and event building blocks (e.g. start, decision) needed to design a Process model. These building blocks do not need corresponding service implementation. The functionality is directly implemented by the Execution engine. The Configurations tree stores several default settings like fallback icons, if a build block does not have a custom icon.

The Tenant specific group is a tree for each tenant, containing the applications, processes, users and roles of this tenant. Aside from general information about applications and Processes, this tree also stores tenant specific meta information like the status of his Process models (e.g. test or productive) or the technical details of his application instances. It is possible for tenants to import their own users and roles into this tree and therefore integrate them in the LPD. This enables tenants to use their roles in the Process model.

The focus of the LPD on Domain specialists (e.g. Logistics) requires a special view on the taxonomy, because the technical view with all public categories can be too detailed and complex for that target group. The LPD offers a simplified modeling view which hides many technical details without affecting the result of an executable Business Process. The user can freely switch between these views.

These are the key features of the LPD:

- *Business Process Execution platform independent*: LPD Processes can be executed on different Process Execution platforms by translating the LPD model into a State machine based Process language. After translation the Processes are executed in the native language of the appropriate Execution platform.
- *Pre-implemented and extensible building blocks*: Every Logistics Mall application offers pre-implemented building blocks for the provided functions and services in a taxonomy. The LPD uses these building blocks directly, without the need for technical refinement. Attached to each building block is an Execution template fragment for the translation into the selected native Execution engine.
- *Data flow based on a standardized Business Object model*: all Logistics Mall applications use the same Business Object model, which is specialized for the Logistics domain. The Business Object model guarantees the interoperability of Logistics Mall applications from different vendors.

The LPD offers several other features to ease the modeling and execution:

- *Multilingual support*: All taxonomy objects can be stored in various languages, enabling the LPD to display information like application descriptions in those languages.
- *Multi-tenant support*: every tenant has access to a tenant specific tree. The tenant-independent part is read-only.
- *Validation of the modeling elements*: the validation of modeling elements is possible because the taxonomy has information about all elements, references and the inheritance relations between them.
- *Compatibility between services*: the compatibility between services can be guaranteed because the interfaces of the services use the standardized Business Objects. So it is easy to check if the output of one service fits the input of another (Fig. 7).
- *Accessibility to users*: the technical view and the simplified modeling view allow all kinds of users to create executable Process models.
- *Automatic code generation*: the usage of Execution templates for activities in combination with template parameters facilitates the automatic code generation for any target platform.

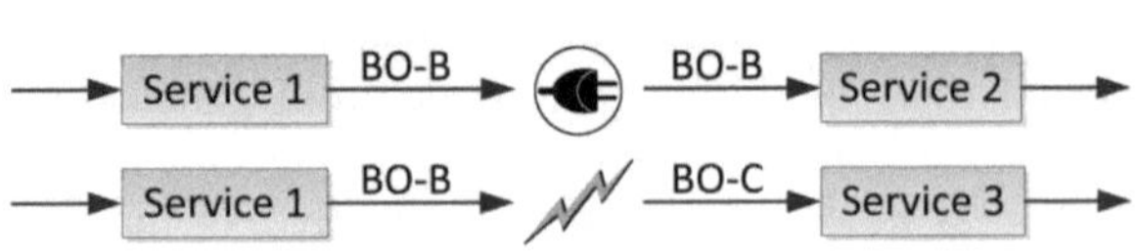

Fig. 7 Compatibility test

- *Versioning*: the versioning of Process models is supported through the taxonomy as a Process model can store its versions in a branch.
- *Flexible modeling notation*: The available Process model elements of the LPD are defined in the PMT, so new model elements can easily be added or the properties (e.g. icons, parameters) of existing ones can be changed without the need for LPD changes.
- *Optimum Performance*: the tree structure of the LPD taxonomy is optimized for read operations and a good performance.

3.2 Modeling of a Basic Example

The Screenshot (Fig. 8) shows the canvas of the LPD with a basic Process model. This process is based on the Nelsan Scenario [17], presented on the *Nationaler IT Gipfel 2012* [18]. The following overview summarizes the used building blocks (yellow rectangles connected with dotted lines are comments) (Table 1):

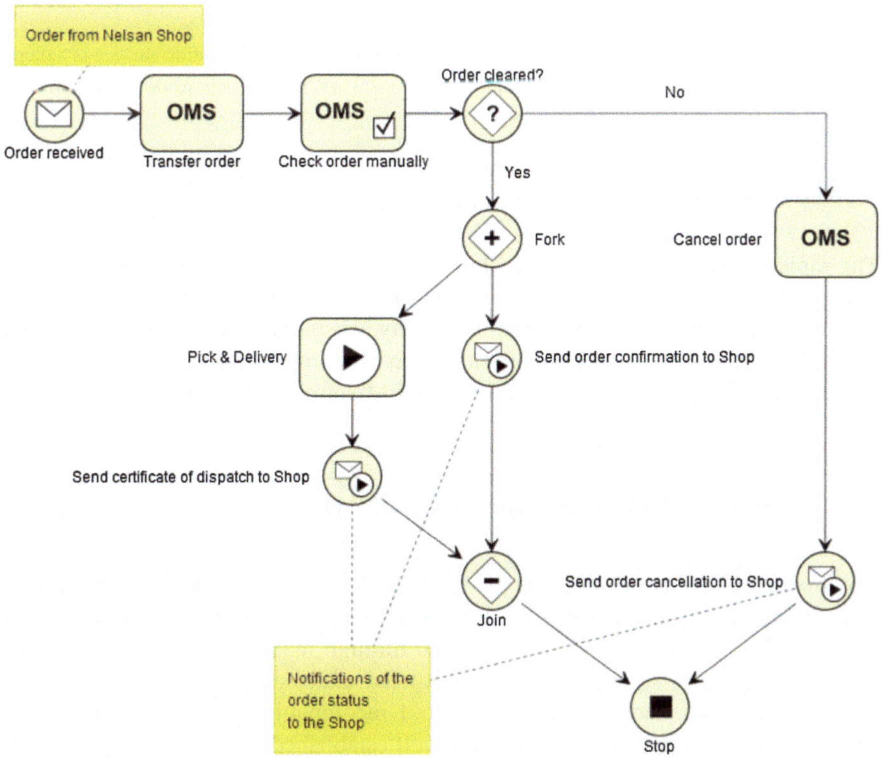

Fig. 8 Order Management Process model

Table 1 Building blocks of the order management process model

Building block	Description
Order received	The process starts with the arrival of a purchase order BO from the online shop.
Transfer order	This activity is an automated service to import the purchase order into the work items of the Order Management System (OMS).
Check order manually	In this human task (visualized by the checkbox in the activity icon) the order is manually reviewed or declined.
Order cleared?	The decision block evaluates the review result and follows either the *Yes* or *No* path.
Fork	If the order was approved, further activities are called simultaneously.
Pick and delivery	This subprocess handles the packing and distribution of the ordered products.
Send certificate of dispatch to shop	After the subprocess is finished, a certificate of dispatch is sent to the shop.
Send order confirmation to shop	The order confirmation is send to the shop.
Join	After both paths are finished, they are joined in one control flow and the process stops.
Stop	Ends this process.
Cancel order	If the order was refused, the activity cancellation service is called.
Send order cancellation to shop	This building block sends a refusal message to the shop and the process stops.

4 Related Work

The rapid development of IT and Cloud based architectures opens up new potentials for Business processes. Especially company overlapping business Processes give room for improvement of the scalability, costs and complexity through the Cloud BPM combination. Many providers have identified those potentials and offer commercial or Open Source Cloud based BPM solutions [19, 20].

4.1 Commercial Vendors for Cloud-BPM

Commercial Cloud BPM products are the majority of available tools. These usually offer a broad lineup of functions for workflows, collaboration, simulation and compatibility. The compatibility is realized through many adapters and connectors for Business Objects. The LPD does not need adapters, as the standardized BO model is used for communication to avoid incompatibilities between interfaces. Many of those commercial products are well established on the BPM market [21]:

Inubit BPM-Suite from Bosch Software Innovations [22] directly generates executable Processes from Process models using customizable templates for the code generation, like the LPD. Inubit offers an own Process engine for the execution of Processes. Existing Processes can be imported from external BPM systems.

Intalio offers the *Intalio|Create* [23] Web based platform which includes an own Process engine. This engine can execute modeled Processes by involving predefined constructors or scripts in various programming languages. Intalio distinguishes between two perspectives, a technical and a domain perspective.

The company *IYOPRO* offers a product with an identical name [24]. The product makes it possible to run Process flow simulations as well as path visualization with automatic validation of executability. The domain Process models can be automated through extensions of the Workflow management. The IYOPRO BPM suite supports collaborative access of Process models and supports various Process engines, including an in-house development, for the execution of those Process models.

The following tools support CBPM only as PaaS solutions, which is not the focus of this article [21]: *Appian BMP Suite* [25], *Cordys Business Operations Platform* [26, 27], *Interstage Business Operations Platform from Fujitsu* [28] and *TIBCO ActiveMatrix BPM* with *TIBCO Silver BPM* [29, 30].

4.2 Open Source Products for Cloud-BPM

Available Open Source solutions in the market supply sufficient functionalities for basic application areas whereas commercial tools outperform with additional features and better support [31]:

Activiti BPM Platform [12] from Activiti is a good representative of this category. This Process engine is used in the Logistics Mall Stage 3. The engine is Javabased with a Model Repository to save Process models. This BPM platform provides the *Activiti Explorer* to view running Process instances. The transition from the domain model to the technical model is only possible with a switch of the used Activiti components: a non-trivial technical refinement of the domain model from the Activiti-Modeler is only possible by using the Activiti Designer. The platform supports test Process execution with the help of unit tests and the task management. The Java code enables the customer to implement his own activities [32].

Intalio/Create is also available in a free community version. In comparison to the commercial version, the *Intalio|Pipes* component is not included. This component orchestrates calls to other systems and components inside the Intalio Cloud.

IYOPRO has also basis version available as Open Source. The essential differences to the commercial version are the missing Process execution and the custom rules for compliance checks.

4.3 Academic Solutions for Cloud-BPM

Only few academic solutions exist, but the growing interest in the subject optimization of Business Processes with connection to the cloud can be ascertained. They are primarily focused on innovative ideas, which are realized in the context of a project and eventually lead to a finished product on the market. The scientific sector provides the following solutions or unfinished projects aside from the LPD.

The Institute of Databases and Information Systems from the University of Ulm presented their Cloud based BPM Solution *clavii BPM Cloud* [33] on the CeBIT 2013. With the help of this solution, every user can work with personalized Process views and make changes to the model. Those changes are immediately published to all other users which are affected by the changes. The personalization is achieved by abstraction of the Business Processes (depending on the field of duty, the necessary activities are shown and the other activities are hidden or grouped). The complexity of Business Processes can be reduced through the selection of various views. Furthermore, it is possible to change a graph based process visualization (BPMN 2.0) in a form based, a textual to an ADEPT[24]-visualization. The clavii BPM Cloud facilitates case based Process changes at runtime. The solution offers a Process modeler and Process execution is planned. The solution emerged from the science project *proView* [34] from the University of Ulm, where the concepts of the Process abstraction were taken from [35].

The Konstanzer Institut für Prozesssteuerung (KIPS) from the HTWG Konstanz [25] works in cooperation with industrial partners on the science project *BPM@Cloud* [36], where the combination possibilities of Cloud Computing and BPM are researched in separate labs (BPMN-Lab, Cloud-Lab and Mobile-Lab). Various Activities which would profit from Cloud Business Processes are determined from the research to optimize the Business Processes based on scientific aspects. KIPS has developed the method $\text{BPM(N)}_{\text{Easy}}$ for that purpose, where the agile software development is combined with the "classical" BPM lifecycle. This method was successfully tested with the introduction of new Processes. The next steps are researching to what extend this method can optimize already running Processes. For tooling the BPM platform *Xpert.ivy* [37] from the Axon Active AG is used. A mobile application for a mobile BPM, which uses Web services to communicate with a Workflow engine running in the Cloud, is in development at the Mobile-Lab. With this application Processes can be created, executed and monitored [19, 36, 38].

---

[24] ADEPT: Application Development Based on Encapsulated Pre-Modeled Process Templates, Process Management System developed from the University Ulm [39].

[25] HTWG Konstanz: University of Applied Sciences.

5 Future Work and Conclusion

In this article we presented the challenges of *Cloud Business Process Management.* A specialized CBPM tool suite can significantly contribute to the optimization and automation of Business processes in the Cloud. We analyzed available tool suites in the market with the outlined requirements and found no adequate solution. Consequently we have developed the Logistics Process Designer including the components and tools shown in Fig. 3. The stage three of the Logistics Mall is our ideal solution for the execution of Business processes in a Cloud for Logistics enterprises.

Main aspect of the LPD is usability and hiding technical aspects: A user does not need either IT knowledge or Process modeling skills. After a quick introduction anybody is able to build a Process model with the LPD.

With the help of pre-implemented building blocks, IT refinement has become unnecessary. The code generator translates a Process model automatically into an appropriate executable model for the integrated Execution engine. Technical details as needed for communication with the outside world are transferred into the Mall gateway. The Process Modeling Taxonomy has a central role in our approach. It contains all available building blocks and defines the corresponding Execution templates. The Cloud operator has to carefully plan and operate the PMT in order to guarantee support for a wide range of use cases to be realized with the solution.

In the future we will complete the LPD with additional features. For monitoring we will use the Process model in the canvas and highlight the current execution and give information about the current Process context. Furthermore, a direct link to the Business Object Instance Repository will give the chance to inspect and modify Business Objects directly from the canvas in case of an exception situation. With a Multi Execution engine support we give the Cloud customer the chance to select an Execution engine most suitable for his enterprise. The code generator will translate the Process model into the correct engine the Cloud customer has selected.

References

1. Mell, P., Grance, T.: The NIST definition of cloud computing, working paper national institute of standards and technology (2009). http://www.nist.gov/itl/cloud/upload/cloud-def-v15.pdf
2. BSA: 2013 BSA Global Cloud COMPUTING Scorecard (2013). http://cloudscorecard.bsa.org/2013/assets/PDFs/BSA_GlobalCloudScorecard2013.pdf. Accessed 29 April 2013
3. OMG: Business Process Model and Notation (BPMN), Version 2.0 (2011). http://www.omg.org/spec/BPMN/2.0. Accessed 29 April 2013
4. Harmuth, N.: Business Process Model and Notation 2.0—Evaluierung der Spezifikation und Ableitung einer formalen Semantik an ausgewählten Sprachelementen auf Basis von Petri-Netzen (2010), Diplomarbeit an der Fakultät für Informatik der TU Dortmund (reference in German)
5. World Wide Web Consortium (W3C): Web Services Architecture. http://www.w3.org/TR/ws-arch/. Accessed 29 April 2013

6. Lawler J.P., Howell-Barber H.: Service-Oriented Architecture: SOA Strategy, Methodology, and Technology. Auerbach Publications, Boca Raton (2008)
7. Holtkamp, B.: The logistics mall—an it-architecture for logistisc-as-a-product. In: ten Hompel, M., Rehof, J., Wolf, O. (eds.) Cloud Computing for Logistics, Lecture Notes in Logistics, Springer (2015)
8. Gsell, H., Nagel R.: Application integration in the logistics mall. Paper presented at the SDPS 2012—17th international conference on transformative science, engineering, and business Innovation. Berlin, Germany (2012)
9. Chappell, D.: Enterprise Service Bus: Theory in Practice, O'Reilly Media Inc., Sebastopol (2004)
10. Böhmer, M., Schmidt, M., Weißenberg, N.: Seamless interoperability in logistics: narrowing the business-it gap by logistics business objects. In: ten Hompel, M., Rehof, J., Wolf, O. (eds.) Cloud Computing for Logistics, Lecture Notes in Logistics, Springer (2015)
11. Wnek, J.: LSI-based taxonomy generation: the taxonomist system. In: Kantor, P. et al. (eds.) ISI 2005, LNCS, vol. 3495, pp. 389–394. Springer, Berlin, Heidelberg (2005)
12. Activiti BPM Platform Homepage. http://www.activiti.org
13. Eggemann, J., Leveling, J., Weiß, N.: Business apps meet the challenge of covering continually changing logistics requirements. In: ten Hompel, M., Rehof, J., Wolf, O. (eds.) Cloud Computing for Logistics, Lecture Notes in Logistics, Springer (2015)
14. Google Developers: Google Web Toolkit. https://developers.google.com/web-toolkit/. Accessed 29 April 2013
15. World Wide Web Consortium (W3C): HTML5 Candidate Recommendation. http://www.w3.org/TR/2012/CR-html5-20121217/. Accessed 29 April 2013
16. yWorks: yFILES for HTML. http://www.yworks.com/de/products_yfileshtml_about.html. Accessed 29 April 2013
17. Bundesministerium für Wirtschaft und Technologie: Exponat „IKT-Innovationen in der Logistik—Wachstumsmotor für Unternehmen", Nationaler IT Gipfel 2012, Retrieved from http://www.post-und-telekommunikation.de/PuT/1Fundus/Dokumente/7._Nationaler_IT-Gipfel_2012_Essen/20_it-gipfel-2012-exponat-ikt-innovationen-in-der-logistik,property=pdf, bereich=itgipfel,sprache=de,rwb=true.pdf (reference in German)
18. Nationaler IT Gipfel Homepage. http://www.it-gipfel.de
19. Mevius, M., Wiedmann, P.: BPM @Cloud-Lab—Open Innovation Platform zur Entwicklung und Evaluation eines agilen Geschäftsprozessmanagements. In: HTWG Forum. Das Forschungsmagazin der Hochschule Konstanz, pp. 60–66. HTWG Konstanz (2012)
20. Mevius, M., Stephan, R., Wiedmann, P.: Innovative approach for agile bPM. In: The Fifth International Conference on Information, Process, and Knowledge Management eKNOW, pp. 160–165. IARIA (2013)
21. Sinur, J., Hill, J.B.: Magic quadrant for business process management suites. In: Gartner RAS Core Research Note G00205212, pp. 1–24. RA2 10182011 (2010)
22. Inubit BPM-Suite Homepage. http://www.bosch-si.com/technology/business-process-management-bpm/business-process-management.html. Accessed 29 April 2013
23. Intalio|Create Homepage. http://www.intalio.com/products/create/challenges. Accessed 29 April 2013
24. IYOPRO Homepage. http://www.iyopro.de. Accessed 29 April 2013
25. Appian BPM Software Homepage. http://www.appian.com/bpm-software/bpm-suite.jsp. Accessed 29 April 2013
26. Cordys Business Operations Platform Homepage. http://www.cordys.com/business_operations_platform. Accessed 29 April 2013
27. Fleming M., Silverstein J.: IDC market scape: worldwide business process platforms 2011 vendor analysis. In: BuyerPulse Analysis: Business Process Management Software, Competitive Analysis, IDC #228598, vol. 1 (2011)
28. Interstage BOP Homepage, http://www.fujitsu.com/global/services/software/interstage/solutions/bop. Accessed 29 April 2013

29. Craggs, S.: Comparing BPM from Pegasystems, IBM and TIBCO. Taking a high-level look at BPM solutions from three leading vendors. In: Whitepaper, 1.00 (2011)
30. TIBCO Software Homepage. http://www.tibco.de. Accessed 29 April 2013
31. Rücker, B.: Open Source BPM. Ein Überblick über aktuelle Tools und Strömungen. In: Javamagazin, DevOps, pp. 86–95 (2012) (reference in German)
32. Preusker, N., Rücker, B.: Activiti 5. Open source BPM on the next level. In: Javamagazin, Social Media API, pp. 106–110 (2010)
33. Clavii BMP Cloud Homepage. http://clavii.com. Accessed 29 April 2013
34. ProView (Personalized and Updatable Process Visualizations) Homepage. http://www.uni-ulm.de/in/iui-dbis/forschung/projekte/proview-project.html. Accessed 29 April 2013
35. Bingmann, A.: Ulmer Informatiker auf der CeBIT 2013: Neue Anwendung garantiert Durchblick bei komplexen Geschäftsprozessen. News. http://www.uni-ulm.de/home/news-details/article/ulmer-informatiker-auf-der-cebit-2013-neue-anwendung-garantiert-durchblick-bei-komplexen-geschaefts.html (reference in German)
36. BPM@Cloud Forschungsprojekt Homepage. http://bpmcloud.in.htwg-konstanz.de/BpmCloud/index.php. Accessed 29 April 2013
37. Xpert.ivy Homepage. http://www.xpertline.ch/en/xpertivy.html. Accessed 29 April 2013
38. Collaboration in der Cloud 3.0—Business Process Management und Social Media. In: Midrange Magazin, ITP Verlag GmbH, online version: (2013). http://www.midrangemagazin.de/mm/artikel.html?id=7992. Accessed 29 April 2013
39. Bauer, T., Dadam, P.: Ifficient distributed workflow management based on variable server assignments. In: Wangler, B., Bergman, L. (eds.): CAiSE 2000, LNCS vol. 1789, pp. 94–109. Springer, Berlin, Heidelberg (2000)

MIX
Papier aus verantwortungsvollen Quellen
Paper from responsible sources
FSC® C105338

If you have any concerns about our products,
you can contact us on
ProductSafety@springernature.com

In case Publisher is established outside the EU,
the EU authorized representative is:
Springer Nature Customer Service Center GmbH
Europaplatz 3, 69115 Heidelberg, Germany

Printed by Libri Plureos GmbH
in Hamburg, Germany